Palgrave Studies in Medieval and Early Modern Medicine

CENTRE FOR THE STUDY OF
MEDICINE AND THE BODY
IN THE RENAISSANCE

The series focuses on the intellectual tradition of western medicine as related to the philosophies, institutions, practices, and technologies that developed throughout the medieval and early modern period (500-1800). Partnered with the Centre for the Study of Medicine and the Body in the Renaissance (CSMBR), it seeks to explore the range of interactions between various conceptualisations of the body, including their import for the arts (e.g. literature, painting, music, dance, and architecture) and the way different medical traditions overlapped and borrowed from each other. The series particularly welcomes contributions from young authors. The editors will consider proposals for single monographs, as well as edited collections and translations/editions of texts, either at a standard length (70-120,000 words) or as Palgrave Pivots (up to 50,000 words).

Michael Stolberg

Medicine and the Body in Early Modern Europe

palgrave
macmillan

Michael Stolberg
University of Würzburg
Würzburg, Germany

ISSN 2524-7387 ISSN 2524-7395 (electronic)
Palgrave Studies in Medieval and Early Modern Medicine
ISBN 978-3-031-84555-0 ISBN 978-3-031-84556-7 (eBook)
https://doi.org/10.1007/978-3-031-84556-7

Cover illustration: Portrait of a surgeon by an unidentified Dutch painter. Metropolitan Museum of Art, New York, Theodore M. Davis Collection, Bequest of Theodore M. Davis, 1915

This Palgrave Macmillan imprint is published by the registered company Springer Nature Switzerland AG
The registered company address is: Gewerbestrasse 11, 6330 Cham, Switzerland

Preface

Over the last two decades, the mass digitization of sources and research literature has fundamentally changed the historian's craft. Many historians now spend much more time every day looking at a computer screen than reading manuscripts, journals, or printed books which they are actually holding in their hands or have lying on the desk before them. Long gone are, for most of us, the days when we had to fill out a paper slip for each title we wanted to see in the library and which we might be able to pick up a week later only. Today, the publications of other scholars in the field are often only a couple of clicks away. As authors, we gratefully experience, in turn, how digitization and the internet have greatly improved our possibilities to reach a wider audience, around the globe.

For many scholars a major barrier remains, however, and it is one that is remarkably little talked about: language. The rise of the English language to the status of the new global *lingua franca* first in the sciences and more recently in the humanities has had undeniable positive effects. Few scholars read more than four or five different languages and many have to make do with even less. In this situation, English-language publications by scholars from non-English speaking countries make their research and findings accessible not only to the sizeable group of native English speakers but also to the much larger number of scholars around the world from other linguistic backgrounds.

The dominance of the English language comes at a considerable price, however, a price of which historians whose mother tongue is not English inevitably tend to be more acutely aware than native speakers. It is a truism, certainly for historians of science and the humanities, that new findings and innovative ideas and theories are not just appreciated and taken up by others due to their intrinsic quality. History has always been and will always be an eminently rhetorical discipline. Stylistic skills and eloquence, the ability to expound arguments and counter objections, to make a case for a certain position, to highlight the importance of one's findings, or quite simply to tell a good story, are crucial for the successful dissemination and reception of a historian's work in the scholarly community. However, even those who, like the author of these lines, had the fortune to live and work for some time in English speaking countries, will almost inevitably never be quite as proficient in English as they are in their mother tongue. In fact, for most of scholars, publishing in English rather than in their own language remains a battle. They have to make a considerable effort to achieve results which, at best, will only come close to the quality they could much more easily and quickly achieve if they used their native language and which often will indeed be clearly inferior, in terms of style, precision, and the ability to make a convincing argument.

The decision to publish this collection of essays results to a considerable degree from my own, personal experience of this dilemma. It reflects the difficult choice I had to make again and again between publishing in German, my mother tongue, in which I feel at home and in which I can best express myself, and, as I have done in several dozen papers, writing in English, a language that is not mine but which allowed me to reach a wider, international readership: the eleven essays in this volume all have already been published elsewhere, but only in German, and have for this reason remained inaccessible to the large majority of scholars outside the German speaking area.

I have selected these essays from a much larger number of major and minor articles which I have written in German over the last thirty years, because I believe they can, in one way or another, still be of value to readers today. Of course, they reflect to some degree the historiographical context, the debates, and major themes at the time of their publication—and as such also my own intellectual journey over all these years. Some of the questions and issues these essays approach have been the subject of extensive further research since then.[1] I myself have drawn on findings

[1] This is true, in particular, for the chapter on patient letters as a source for the historical experience of illnesses and the body (see, for example, Olivia Weisser, *Ill composed. Sickness,*

presented in these essays in later publications and especially in my books.[2] What all the papers have in common, however, is that they draw heavily on my study of a wide range of Latin and vernacular sources, which have not been studied in similar depth—or indeed not at all—by others since then. This, I would hope, makes them to some degree timeless.

The essays are arranged in a rough thematic order. The volume starts with several papers on various aspects of the perception and experience of the healthy and sick body in the early modern period, such as the then widespread and nowadays largely forgotten concept of critical "climacteric" years in men and women which eventually morphed into the modern notion of the female climacteric, learned and lay beliefs about the pathogenic effects of strong emotions and the lay experience of illness as expressed in patient letters addressed to famous physicians like Samuel Auguste Tissot. I offer a new interpretation of Lukas Cranach's famous "Melancholia"-paintings in the light of Reformed anthropology and the often overlooked concept of melancholy as a disease rather than a temperament, and I sketch the rise of the concept of "palliative care" since the late Middle Ages. The following essays focus on the learned physicians. I discuss their self-fashioning, the challenges to their professional authority they constantly faced, due not least to the frequent failure of their treatment efforts, and the different strategies to which they resorted to assert their authority. Two further essays offer a typology of the forms and functions of physicians' casuistic writings in the early modern period

gender, and belief in early modern England, New York and London 2015) and for my study of early modern medical case histories and *observations*, a genre, to which Gianna Pomata, in particular, has since then devoted various papers (see Gianna Pomata, "Sharing cases: The *observationes* in early modern medicine", in: *Early science and medicine* 15 (2010), pp. 193–236; eadem, "A word of the empirics: The ancient concept of observation and its recovery in early modern medicine", in: *Annals of science* 65 (2011), pp. 1–25; eadem, "Observation rising. Birth of an epistemic genre, 1500–1600", in: *Histories of scientific observation*, ed. by Lorraine Daston and Elizabeth Lunbeck, Chicago and London 2011, pp. 45–80; see also Katharine Park, "Observations in the margins, 500–1500", in: ibid., pp. 15–44). The history of emotions even has developed into a major area of historical inquiry over the last decades although the physiological and pathological concepts that guided specifically their medical understanding remain to be studied in greater detail (see also my more recent paper on "Emotions and the body in early modern medicine", in: Emotion review 11.2 (https://doi.org/10.1177/1754073918765674).

[2] See, in particular, my *Experiencing illness and the sick body in early modern Europe* (2011), *History of palliative care, 1500–1970. Concepts, practices and ethical Challenges* (2017), and *Learned physicians and everyday medical practice in the Renaissance* (2022).

and discuss the Paduan *collegia*, the public discussion of case histories by various professors, which served as a fruitful didactic tool but were also a forum in which professorial hierarchies were played out. The volume concludes with a two doxographic papers which, based on a close reading of the relevant Latin sources, discuss the place and development of some of the key concepts of late medieval and early modern physiology and pathology—the innate heat and its relationship to the spirits, diseases and drug effects of the "total substance" and their relationship to the notion of "form" in the work of Avicenna and Daniel Sennert.

With the exception of this introduction and the footnotes, the texts in this volume were translated into American English by Logan Kennedy and Leonhard Unglaub, with whom I have been working for many years. I have left the essays unchanged otherwise, refraining from revising them in the light of more recent research or from even just offering references to more recent publications that would seem relevant for the topic or issue in question.

I would like to dedicate this volume to the memory of Eckhard Keßler, with whom I studied at the University of Munich. He was not only a leading expert on Renaissance philosophy but also an inspiring teacher and decisive for making me go back in time, from my initial area of research—health care, public health, cholera, and environmental issues in the early nineteenth century—to the even more foreign world of the early modern period. It has been an enjoyable and rewarding journey to this day.

Würzburg, Germany Michael Stolberg

Contents

List of Figures

From "Step Years" to "Menopause". The Changing Notion of the "Climacteric"

The Doctrine of the "Climacteric Years"

"Every seven years there is a climacteric year in which humans are subject to mortal danger. And when seven times nine comes, it is the sixty-third year, and this is the greatest year of danger, which breaks the necks of some, or to speak more plainly, closes their eyes for good". With these words, written in 1734, a *Lustiger Philosophus* ("Merry Philosopher") presented to his readers a doctrine which seems outlandish to us today but which by all appearances was widespread and recognized in earlier centuries. It was the doctrine of the *anni scalares,* the climacteric years, or *anni climacterici,* a term whose Greek root refers to "steps" or "rungs of a ladder".

The origins of this doctrine go back to antiquity. As the *Philosophus* correctly explained to his readers, each person's body and mind went through a marked change every seven years (though, for some authors, it was every nine years). In their seventh year, children lost their baby

---

First published in German as "Von den 'Stufenjahren' zur 'Menopause'. Das 'Klimakterium' im Wandel der Zeit", in: *Würzburger medizinhistorische Mitteilungen* 26 (2005), pp. 41–50 Translated into American English by Logan Kennedy and Leonhard Unglaub.

M. Stolberg, *Medicine and the Body in Early Modern Europe*, Palgrave Studies in Medieval and Early Modern Medicine,
https://doi.org/10.1007/978-3-031-84556-7_1

teeth and their permanent teeth came out. The fourteenth year marked the beginning of puberty. In their twenty-first year, the beard sprouted and reason grew, and this seven-year pattern of growth was more or less how things went for the rest of the person's life. At the same time, grave danger loomed in every seventh or, in the less widespread enneadic conception, every ninth year as well, putting life and limb in jeopardy because in these years "diseases and other accidents generally tend to make themselves known". The most dangerous of all years was the "grand climacteric", the sixty-third year, in which the cycles of seven and nine converged and in which, as the *Lustiger Philosophus* wrote, "countless people, of both noble and lower standing" had lost their lives. Hardly less dangerous were, however, the forty-ninth year (seven times seven), also known as the "little climacteric year", the eighty-first year (nine times nine), and the fifty-sixth year (seven times eight), which was also given the moniker *climacter heroicus,* or the "step year" of the nobles.[1]

This old idea of "climacteric" years has sunk almost entirely into oblivion. Even most medical historians are no longer familiar with it. Yet in pre-modern times, it apparently had a significant influence on people's experience of the course of human life and its cycles, and in the sixteenth and seventeenth centuries in particular, it moved the minds of learned scholars. Dozens of contemporary writings addressed the significance of the "climacteric years". Some of them were devoted exclusively to this subject, including the almost 900-page work *De annis climactericis* by Claudius Salmasius (Claude Saumaise) of 1648.[2] Most of these works described the "climacteric years" as something indisputably proven by experience. Almost everybody could name numerous examples of famous and less famous people who had died in a climacteric year, especially in the *annus climactericus maximus,* the sixty-third year of life. Some authors filled dozens of pages with lists of people who died in a climacteric year.[3] Even the proverbial longevity of the ancestors in the bible served as proof.

[1] *Der Lustige Philosophus, Welcher Mancherley lächerliche und ergetzende Fragen und Antworten, Gespräche und Distinctiones, so wol in die Philosophie, als Jurisprudentz und Medicin lauffend, vorbringet* [...], Frankfurt and Leipzig 1734, pp. 290–1.

[2] Claudius Salmasius, *De annis climactericis et antiqua astrologia diatribae*, Leiden 1648.

[3] Henricus Ranzovius, *Exempla quibus astrologiae scientiae certitudo,doctissimorum cum veterum, tum recentiorum auctoritate astruitur [...]. De annis climactericis*, Cologne 1585.

As was well-known, Adam had lived to the age of 930 years. That meant that he died in his 931st year, and 931 was 133 times seven.[4] In 1623, Baptista Codronchi found further powerful evidence in the parish registers. The death records, he claimed, showed that every fourth death in people over sixty took place precisely in the sixty-third year, and of the remaining deaths, only a few took place outside one of the later "climacteric years".[5] Supported by arguments such as these, most of the authors of the day were in agreement that the doctrine of the "climacteric years" was, as Ranzovius put it, "not superstitious, but born of custom and experience".[6] It was only later, in the eighteenth century, that Süßmilch and other proponents of the new field of demographic statistics opposed this "erroneous judgement" with their data. They explained that while a great many people did die in their sixty-third year, the risk of dying was even greater in the sixty-fourth year and it continued to grow in the sixty-fifth year.[7] The skepticism corresponded with an overall trend: in the eighteenth century, the old doctrine of the "climacteric years" became progressively marginalized in learned discourse. It would continue to find frequent mention until early in the twentieth century, but authors increasingly wrote about it from a historical perspective.

Disputing Causation

While, in the sixteenth and seventeenth centuries, there was still general agreement about the significance of the "climacteric years", the controversies about how to explain the changes and dangers of these years were all the fiercer. Learned authors largely rejected the numerological mysticism that was likely at the root of the belief, namely the idea that numbers in themselves might have a special power. They also rejected the belief that the number seven was particularly pleasing to God. It had come to be

4 Baptista Codronchi, *De annis climactericis:Necnon de ratione vitandi eorum pericula,itemque de modis vitam producendi commentarius*, Cologne 1623, p. 10.

5 Ibid., p. 21.

6 Henricus Ranzovius, *De conservanda valetudine liber*, 3rd edn, Antwerp 1584, p. 100.

7 Johann Peter Süßmilch, *Die göttliche Ordnung in den Veränderungen des menschlichen Geschlechts, aus der Geburt, dem Tode und der Fortpflanzung desselben erwiesen*, Berlin 1761, pp. 361–363; similarly already Henricus Löchstör, *Dissertatio physica de vano metu anni climacterici magni sexagesimi tertii*. Def. Petrus Nicolaus Kleist, Copenhagen 1737, p. 14.

commonly understood that numbers and measurements of quantities as such did not have any effects. They were mere abstractions.[8]

Astrological interpretations met with greater approval. Astrologers ascribed the drastic changes and dangers that accompanied every seventh year of life to the action of the planets. Particular attention was paid to Saturn, known for its malignant influence, whose movement through the celestial quadrants followed a seven-year cycle.[9]

To the extent that the danger of the climacteric years was considered an established empirical fact, it conversely offered the adherents of astrology welcome empirical evidence of the significance of their teachings and of their expertise. Among the scholars of the sixteenth and seventeenth centuries, however, the belief in a direct influence of the planets on earthly affairs increasingly met with criticism, and by 1700, astrological beliefs were held by only a small minority.[10] In the process, the astrological interpretation of the "climacteric years" was largely rendered obsolete.[11] There were objections on astronomical grounds as well, for example the observation that the (apparent) orbit of Saturn around the Earth took not twenty-eight years but approximately twenty-nine and a half.[12]

[8] Ambrosius Floridus, *Tractatus de annis climatericis [sic!], ac diebus criticis. Dialogistico contextus sermone*, Padua 1612, p. 6; Codronchi, De annis [see note 4], p. 7.

[9] Marsilio Ficino, *Three books on life. A critical edition and translation with introduction and notes.* Ed. by Carol V. Kaske and John R. Clark. Binghamton, NY, 1989, p. 230: "itaque in septimo quolibet vitae anno fit in corpore mutatio maxima ideoque periculosissima, quandoquidem et saturnus nobis communiter est peregrinus, et ab eo tunc planetarum summo ad lunam e vestigio planetarum infimam gubernatio redit." Cf. Floridus, Tractatus [see note 8].

[10] Wolf-Dieter Müller-Jahncke, *Astrologisch-magische Theorie und Praxis in der Heilkunde der frühen Neuzeit*, Stuttgart 1985, pp. 135–175.

[11] See the detailed critique in Ioannis Baptista Sylvaticus, *De anno climacterico tractatus*, Pavia 1615; Salmasius, De annis [see note 2]; Joachimus Leistenius, *Pro felici fluxu anni climacterici magni,quem reverendissimus et serenissimus princeps Dn. Augustus primas Germaniae et dux Saxoniae, Juliae, Cliviae, et Montium, &c. &c. conditor*, Weissenfels 1676; Christophorus Heinricus Löber, *Superstitio climacterica, Das ist Gründliche Widerlegung des sehr gemeinen Aberglaubens von den so genannten Bösen Stuffen-Jahren*, Jena 1697.

[12] Joachimus Leistenius, *Reverendissimo,serenissimoque principi Dn. Augusto Primati Germaniae, et duci Saxoniae &c [...] de climacterico magno feliciter superato humilime [sic!] gratulatur*, Weissenfels 1677 (no page numbers).

In place of these beliefs, the conception ultimately took hold across wide reaches of society that the seven-year rhythm was grounded in the laws of nature and in the laws of the human body itself. To counter skepticism, reference was made to the different, similarly regular, periodic processes in the human body whose significance was undisputed, for example the old doctrine of the "critical days". These were certain days in the course of an illness—including the seventh—on which the patient's condition took a turn for better or for worse, much in the way that the "climacteric years" in a person's life introduced a new phase of growth and maturation or else abruptly ended life.[13] Female menstruation, too, followed its own rhythm that was determined by nature: four weeks, that is four times seven days. In older medicine at least,[14] menstruation was interpreted much in the way that attacks of sweating or other forms of excretion on the "critical" days of an illness were interpreted, as "critical" excretions that freed the female body of harmful matter. The Hippocratic theorem, too, according to which children born after eight months of pregnancy were not viable, while those born after seven or nine months survived, continued to be taken seriously to a large extent. Physicians and natural philosophers had to accept such temporal regularities, like the various cyclical changes in inanimate nature, as empirical facts, as given by God or Nature—just as one could observe gravity by way of its effects alone, without having a causal explanation for it.[15]

Some physicians, however, attempted to interpret "climacteric" events in a more concrete way, within the framework of the dominant medical theories, in humoral-pathological terms. They held that the body increasingly filled with corrupt humors in periods of seven years, until the person in the seventh year either excreted them or else fell ill due to their effects, especially when the body was weakened from other influences. The *Lustiger Philosophus* for example described the dangers that were particularly imminent for the learned in their fifty-sixth year, writing

[13] Löchstör [see note 7], pp. 3–4, even considered the notion of "critical days" an important source for the belief in "climacterical years".

[14] Michael Stolberg, "'Zorn, Wein und Weiber verderben unsere Leiber'. Krankheit und Affekt in der frühneuzeitlichen Medizin", in: *Passion, Affekt und Leidenschaft*, Wolfenbüttel 2005, pp. 1033–59 [English translation in this volume].

[15] Karl Ludwig Funckert, *Dissertatio physico-medica inauguralis,sistens annorum climactericorum rationalem et medicam explicationem*. Praes. Friedrich Hoffmann, Halle and Magdeburg 1704.

that at this time “a pathological humor is present with many, and they are subject to many an emotion so that, with all these complaints, some easily bid their farewell”.[16] Pointing out additional triggering factors such as emotions and dietary errors, it could further be explained why some of the people born in a given year did not fall victim to a specific “climacteric year”. The question remained unanswered, however, as to why the marked individual differences between people did not produce a stronger variation in the periodicity, i.e. why certain people did not go through a “climacteric year” every six years, for example, or only every eight years.

Another possible explanation was espoused especially by those authors who were fundamentally skeptical about the doctrine of the “climacteric years” but who did not want to dispense entirely with the established experience of increased mortality in the “climacteric” sixty-third year and perhaps other years as well. For them, the true cause was the great fear that people felt about this year, because fear, like other strong emotions, was considered one of the very most important causes of illness.[17] The heightened mortality in the “grand climacteric” was, in other words, ultimately the work of those who had incorrectly claimed that such an increased mortality was a law of nature, as it were. After all, as Hilscher asked in 1747, what was more frightening to people than death? The great and constant fear that people felt as they approached their sixty-third year threw the humors and vital spirits into turmoil. That alone, or in combination with an unhealthy diet or adverse weather conditions, led to the worst and deadliest diseases.[18]

The “Climacteric Years” in Lay Culture

In retrospect, only a rough idea can be gained of the precise extent to which the doctrine of the dangers of the “climacteric years” was widespread among the population and was a part of everyday life. Printed

[16] Lustiger Philosophus [see note 1], p. 291.

[17] Cf. Funckert, Dissertatio [see note 15], p. 17: “quando existimant, hunc numerum esse periculosum & imminere malum corpori, inquieto & perturbato sunt animo, timent, tristantur, anguntur, idemque periculum semper ipsis imminere cogitant”; cf. Stolberg, Zorn [see note 14] on the importance of the belief in the pathogenic effects of the emotions in early modern medical culture.

[18] Simon Paul Hilscher, *Prolusio de vano aetatis humanae anni LXIII, climacterici magni vulgo dicti,timore*, Jena 1747, p. 6.

speeches and birthday eulogies show, however, that the teaching had at least some presence in the wider population insofar as these texts highlighted the “climacteric year” a ruler or another high-ranking person had just survived or was approaching.[19] But not only this: educated contemporaries made casual reference to their “climacteric years” in other contexts as well and mentioned them in their private correspondence. The term was also used in literary titles.[20] Physicians congratulated one another for having gotten through a “climacteric year”.[21] One might still want to see all this as a literary topos first and foremost, but this interpretation collapses when patients (or their relatives) anxiously inquired about whether the current complaints might have to do with the “climacteric year” in which they found themselves, or when contemporaries attributed the illnesses or deaths of their relatives, in retrospect, to the “climacteric year” of the patient.[22]

We lack relevant ego-documents from the wider, less-educated population. But here too, as can be gathered from the descriptions of physicians and other learned men, the idea was widespread. In 1615, Sylvaticus, for example, described the notion as being “long established and stubbornly received and held on to in the opinion of the people [vulgus]”.[23] Similarly, Superintendent Löber called this opinion “very common” in 1697. According to him, it was “still today rooted with many, including learned and pious people, to such a degree [...] that they think of it as an article of faith”. He wrote that his professor had once taken him to the churchyard expressly in order to show him the gravestones and thus illustrate the dangers of the “climacteric years”.[24] And as late as 1761, Süßmilch

[19] Leistenius even published such a eulogy at the beginning as well as at the end of the “great climacteric year” of his prince (Leistenius [see notes 10 and 11]).

[20] Franz Callenbach, *Eclipses politico-morales. Sicht- und unsichtbare sittliche Staats-Finsternussen, Denen Kindern des Liechts zur Warnung, denen Kindern dieser Welt (die sich klüger düncken als jene) zur heilsamen Verschämung treumeynend vorgestellt [...] Ad annum climactericum magnum praedominante Saturno, die critico deprehensae,tubo scenico perlustratae*, Umbria [1714].

[21] Burgerbibliothek Bern, Ms. 495, Nr. 286, copy of a letter from Gregor Horst to Wilhelm Fabricius, October 1623.

[22] Michael Stolberg, “Zeit und Leib in der medikalen Kultur der Frühen Neuzeit”, in: *Autorität und Zeit in der Frühen Neuzeit*, ed. by Winfried Schulze, Ralf-Peter Fuchs, et alii, Münster 2006, pp. 49–68.

[23] Sylvaticus, De annis [see note 11], dedicatory letter to the senate in Milan.

[24] Löber, Superstitio [see note 11], frontispiece, p. 12, and pp. 9–10.

complained that there were many people "including otherwise reasonable people, who allow themselves to be frightened by this view, and who then feel calm again once they get over this supposed hill".[25]

If all of this is true, then we are dealing here with far more than another small brick in the complex theoretical edifice of early modern medicine. Rather, we would have to understand life as a whole, including life as it unfolded in the everyday subjective experience of ordinary people, as having been marked by a very different, more cyclical dimension. The "climacteric years" would have been the temporal vanishing point of every seven-year period of life—a step on which you could slip and fall or which, if you successfully overcame it, led to the next phase of your life which you met with a renewed, strengthened body. This conception of a seven-year cycle focused on the physical, the bodily aspects only, as opposed to the well-known depictions of a staircase or steps of age divided into decades. The latter also showed physical changes but were particularly concerned with the changes in social status and thus often pictured men at the peak of their lives at the age of fifty.[26]

"Climacteric Years" and the Modern "Climacteric"

Today, the terms "climacterium" and "climacteric" are known and used far beyond the medical profession. Obviously, the meaning of these terms today is completely different from the meaning they held in the old doctrine of the "climacteric years". Let us first emphasize the most important commonality: the "climacteric" in today's understanding continues to denote a critical time of change for the most part, a time of marked physical and psychological transformation. In terms such as "change of life" or *Wechseljahre* in German, which are commonly used as synonyms today, this sense resonates all the more.

However, the differences between the old idea of the "climacteric years" and the modern-day "climacteric" are much more striking. Three of them in particular should be underscored:

[25] Süßmilch, Göttliche Ordnung [see note 7], p. 361.

[26] See the richly illustrated exhibition catalogue *Die Lebenstreppe. Bilder der menschlichen Lebensalter*, Cologne 1983.

1. Today, the climacteric is largely associated with the female body. While there is an ongoing discussion about the existence of a "male climacteric",[27] it remains a matter of dispute and hardly anyone would ascribe to it the same prevalence and significance that is ascribed to the female climacteric. By contrast, the old concept of the "climacteric years" was not associated with the female sex. Quite the contrary: the relevant texts almost exclusively refer to men, and the historical figures who are mentioned were almost always male, as were most famous figures in general in the old male-dominated societies.
2. The "climacteric" in today's sense of the word refers to something that happens once in a lifetime; a seven-year cycle does not come into it. At most, one could see a remnant of the old concept in the fact that the "climacteric" is thought to begin at around the age of forty-nine, that is, in the year that used to be thought of as the "little climacteric year". A similar age—the fiftieth year of one's life—was traditionally understood to mark the beginning of (early) old age, however.[28]
3. The "climacteric" today is not limited to a single year. The term describes a stage of life whose duration varies but is generally quite long, and whose beginning and end can only be loosely determined.

In the light of these distinct differences, the question arises as to how this fundamental shift in meaning could have taken place. Answering it proves to be difficult, not least of all because this shift occurred in a time when terms like "climacteric" and "climacterium" were rarely used at all.

To start with, the sources contradict each other and make it impossible to determine with certainty the exact time when the meaning began to shift, let alone the reasons. As early as 1840, a forensic physician called Adelmann from Gerolzhofen complained about colleagues who began to "twist the meaning of words that have been hallowed by antiquity". His prime example was the German term "climacterisch", which he said people used in "recent times" to speak of the "involution period of the

[27] Wolfgang Schuhmayer, "Die Andropause: Mythos oder Wirklichkeit?" in: *Orgyn. Das Organon Magazin Frauen & Gesundheit* 8 (1997), pp. 15–18.

[28] Manfred Welti, "Das Altern im Mittelalter und in der Frühen Neuzeit", in: *Schweizer Zeitschrift für Geschichte* 37 (1987), pp. 1–32.

female sex [...] as often happens in magazines".[29] In Adelmann's view, this was wrong: "The climacteric years are common to both sexes, so it is wrong to apply the name 'climacteric years' to the involution period of the female sex exclusively".[30]

Adelmann's complaint would seem to allow for a fairly precise dating of the shift in meaning. However, looking more closely at contemporary medical encyclopedias and German medical publications of the time, Adelmann's finding is only corroborated to a small extent: a review of numerous volumes of periodicals shows that the term "climacteric" only begins to appear more frequently after about 1860, and that it was not until the end of the nineteenth century that the term became more common and was now, at least in the medical field, being used in its new, modern sense. The *Enyclopaedia Britannica* of 1910, for example, defined the climacteric as "a critical period in human life", adding "in a medical sense, the period known as the 'change of life', marked in women by the menopause".[31] All that can be said with certainty is that sometime in the course of the nineteenth century, the term "climacteric" came to be primarily applied to the menopausal time in a woman's life.

The fact that such a shift in meaning was possible at all seems to have been due to developments in another medical field, namely in the conception of "menopause", the age-related cessation of female menstruation, and its consequences for a woman's health and life. To understand this better, we need to take a step back again into the early modern period.

From "Cessatio Mensium" to "Menopause"

Beginning as early as the sixteenth century, terms such as "cessatio mensium" were used to describe the age-related cessation of menstruation as a source of many illnesses. In 1563, for example, the well-known Italian physician Giovanni Marinello wrote that just like women whose menstruation ceased for other reasons, those women who stopped menstruating due to age were "always sickly, and usually in those parts of the body that are connected to the uterus or have a certain relationship with it, such as

[29] Dr. Adelmann, "Etwas über das Wort 'Climacterisch'", in: *Medicinisches Correspondenz-Blatt bayerischer Aerzte* 1 (1840), pp. 12–14 and pp. 20–21.

[30] Ibid., p. 21.

[31] The Encyclopedia Britannica, 11th edn, vol. 6: Châtelet to Constantine, Cambridge 1910, p. 509.

the stomach or the head. As soon as menstruation stops in other words, there come pain, ulcers, eye complaints, weak sight, vomiting, and fever. Soon there are tightness in the chest, a fainting heart, shortness of breath, hiccups and other pernicious complaints".[32] Numerous descriptions of a similar nature penned by other physicians can be found dating from the seventeenth and eighteenth centuries.[33] In particular, uterine cancer—the disease that contemporaries considered to be the most harrowing of all illnesses—was attributed to the cessation of menstruation. As experience taught, the disease befell women of this age primarily, just as cancer as a whole was considered a female condition first and foremost.

We can better understand these accounts by looking at traditional humoralism. According to the traditional view, menstruation was a "monthly purification". It freed the female body of the diverse spoiled, impure matter that accumulated in it over the course of the month and thus it was virtually indispensable to health and the preservation of a woman's life. Obviously, great danger therefore lurked when menstruation stopped due to age or disease. The putrid, spoiled matter continued to build up in the body, wreaking all kinds of havoc or releasing harmful, pathogenic fumes and vapors. The body had to try to rid itself of the spoiled matter in other ways, through other excretions, for example by bleeding from another body part. However, if this proved unsuccessful, disease followed almost inevitably.

As is shown by numerous patient letters, many women until well into the nineteenth century were frightened when their menses were merely diminished or delayed. Worried they would turn to a surgeon or a physician in the hope that bloodletting, purgatives, or a remedy to "purify the blood" would free them from the dangerous humors and avert the danger. For many women, reaching the age at which their periods would cease accordingly raised fears, and, as late as the mid-nineteenth century, quite a few were said to take "menstruation-promoting" or "blood purifying" remedies at the first sign that their menstrual periods were ending.

32 Giovanni Marinello, *Le medicine partenenti alle infermità delle donne*, Venice 1563, fol. 87v.

33 On the following see my detailed analysis in Michael Stolberg, "A woman's hell? Medical perceptions of menopause in preindustrial Europe", in: *Bulletin of the history of medicine* 73 (1999), pp. 408–28.

The large majority of medical authors had already parted company with the old understanding of menstruation as purification around 1600. They now described menstrual blood as pure and nutritious, serving primarily to provide nourishment to the fetus. The only reason it was discharged when conception did not occur was that it would have otherwise kept accumulating in the body. However, in this new understanding, the old conviction that it was harmful to miss periods persisted, even though the detrimental effects now had to be explained differently. For example, it was said that these effects arose from stagnating blood in the overly filled and strained vessels, or that a secondary putrefaction occurred due to the extended time during which the matter remained motionless in the lower abdomen. Countless observations of women who fell ill more or less immediately after their periods stopped rendered the deleteriousness of *suppressio* or *cessatio mensium* an incontrovertible fact of experience.

According to all of these interpretations, in other words—and this is important to our inquiry—the age-related cessation of menstruation did not point to a "critical" period in the traditional sense. What was missing was the "critical" excretion as found in the traditional concept. And furthermore, the age-related cessation of menstruation, considered as a period of crisis that a woman survived, did not result in a renewed, strengthened body. On the contrary: when their menstrual cycles stopped, women had to reckon with the fact that matter was going to accumulate more and more in the body, and that the aging female body could at most attain a certain precarious equilibrium so long as other routes of excretion assumed the role that menstruation had played, or—according to later medical conceptions—as long as the production of surplus nutritional blood gradually slowed. However, the primary causes of the cessation of menstruation were generally seen either in the advancing weakness of the female nature, which was no longer able to rid the body of impure matter, or else as in an increasing hardening of the vessels and fibers, which made excretion increasingly difficult on the one hand and on the other gave rise to dangerous bleeding in cases where one of the hardened vessels burst and was no longer able to contract on its own. None of these possibilities augured well for women, neither in the short nor in the long term.

It is only in the late eighteenth century that we come upon the signs of a fundamental shift in the medical discourse, prompted above all by new general conceptions of the body. Influenced by vitalism, physicians began to conceive of menstruation as a specific active secretion via the uterus, analogous to the working of the glands. With this new view, the supposed health-promoting, life-preserving functions of menstruation receded to the background. At the same time, the negative accompanying symptoms—the manifold complaints women experienced immediately before or during their periods—attracted increasing attention. In earlier times, menstruation had been referred to not only as a "blessing of nature" but also as a female "disease". And now this pathological, pathogenic aspect of menstruation was being increasingly emphasized. It was said that the uterus and with it the entire nervous system and the female organism as a whole were put in a state of extreme tension and irritation every month. And so, menstruation was seen as a massive burden for the woman, a great sacrifice, a tribute that she had to pay for her indispensable contribution to procreation.[34]

Almost by necessity, this negative assessment of menstruation brought with it a different and more positive assessment of the age-related cessation of menses. Finally, women were no longer subject to the downright pathological state of aggravation or to the multiple complaints that generally accompanied menstruation. They were now living for themselves, for their own wellbeing, and no longer for that of mankind. Writing from experience, many medical authors explained that older women who had stopped menstruating often were even much healthier than men of the same age and, according to the mortality statistics, also lived longer. Many women even experienced a second youth, were full of energy, and profited from having been released from the burdens of their fertile existence.

At the same time, medical authors from the end of the eighteenth century onward, and at first particularly in France, also described a specific phase of transition from women's fertile to infertile time. This was the phase during which the female periods became more irregular and ultimately stopped, and in which the "sexual function" of the woman waned. This phase they regarded indeed as one that was accompanied by manifold dangers, because in this time the entire female organism underwent a transformation—literally a revolution. The rule of the uterus over the female body was coming to an end, but often not without a final rearing up of the old tyrant. For this reason, women of around forty-five years of age went through what was widely termed an *âge critique*, a critical age that spanned several years. To describe this *âge critique*, this *maladie de la femme de quaranteans,* the young French physician Gardanne

[34] Pierre Roussel, *Système physique et moral de la femme*, Paris 1775, p. 197.

coined a new term in 1812: *ménépausie*,[35] which was soon changed to *ménopause*.[36] In a literal sense—as the physicians at the time with their training in the classical languages knew—*ménopause* simply meant a "stopping" or "ceasing" of the monthly period. Gardanne, however, used it to denote the entire physical process and the diverse symptoms suffered by women in this phase of life.

Only gradually did the term "menopause" find wider use, but the idea of a more or less specific critical phase of life for women, which was overshadowed by the danger of illness but could result in a renewed, stronger, perhaps even more youthful body if all went well, reflected the predominant medical conception already in the decades before Gardanne's publication. In this view, the age-related cessation of menstruation was no longer attended by more or less constant and inescapable physical decline, but was rather the most prominent feature of a prolonged critical transition of the entire organism. This transition occasionally ended in death, but the physicians reassuringly promised that, if one took good care and had some medical advice, it led the way to a period of renewed, strengthened health.

In this way the interpretation of "menopause" came closer in one central point to the traditional understanding of the "climacteric" years: menopause became a distinct phase of life characterized by change and transition. At almost exactly the same time that Gardanne coined the term *ménopausie,* the term "climacteric" also attained its own new connotations. In 1813, in a work that would be printed and cited numerous times, the British physician Henry Halford, writing about the so-called "climacteric disease", presented the public with a new disease entity that he claimed to have encountered again and again in his practice. He counted among the typical symptoms of general exhaustion and fatigability, emaciation, insomnia, and an accelerated pulse without the presence of complaints that could be attributed to a specific location in the body. The condition appeared between the fiftieth and seventy-fifth year of life and was not to be confused with the natural aging processes: it was a

[35] Charles P. D. de Gardanne, *Dissertation sur les avis à donner aux femmes qui entrent dans l'âge critique*, Paris 1812.

[36] Charles P. D. de Gardanne, *De la ménopause,ou de l'âge critique des femmes*, Paris 1821.

temporary pathological state that did not affect all people and of which some patients recovered entirely.[37]

Halford found the "climacteric disease" in both sexes and even more commonly in men but he applied the term "climacteric" to a single period in life, which lasted far longer than a year, as in the new conception of the cessation of menstruation and its effects that emerged around the same time. From this perspective, the differences between Halford's "climacteric disease" and the condition of "menopause" were only a matter of degree. Because, as experience taught, it was predominantly women who suffered the typical complaints at the age in question, a conflation of the two terms was palpably close.[38] It remains for further research to discover more specifically how and why the merging of the terms took place and how the new understanding of the terminology gradually took hold in everyday language.

[37] Henry Halford, "On the climacteric diseases", in: *Medical transactions of the Royal College of Physicians in London* 4 (1813), pp. 1–15.

[38] Eventually, in the twentieth century, when the knowledge of ancient Greek became less and less common, the term "male menopause" was coined and came to be used, although the term makes no sense from an etymological point of view, since "menopause" literally means the end of menstruation.

"Wrath, Women and Wine, Throw Our Bodies to the Swine": Affects and Illness in the Early Modern Period

The affects or passions played an important role in early modern medicine.[1] As phenomena that were experienced or observed at the borderline of body and soul every day, they were the subject of theoretical debate and controversy. When applying the doctrine of the temperaments or making medical diagnoses, it was deemed essential that they be given full consideration because an individual's inclination toward certain affects was thought to provide helpful clues about the person's particular constitution and temperament,[2] and therefore his or her susceptibility to certain

First published in German as "'Zorn, Wein und Weiber verderben unsere Leiber'. Krankheit und Affekt in der frühneuzeitlichen Medizin", in: *Passion, Affekt und Leidenschaft in der Frühen Neuzeit*, ed. by Johann Anselm Steiger and Ralf Georg Bogner, Wiesbaden 2005, pp. 1033–1059. Translated into American English by Logan Kennedy and Leonhard Unglaub.

[1] Following contemporary usage, I will use the terms "affectus" and "passiones" synonymously. The distinction between lasting "passions" and sudden, brief "affects" is of a more recent date; see, for example, the entry "Affect" in J. S. Ersch and J. G. Gruber (eds), *Allgemeine Encyclopädie der Wissenschaften und Künste*, Leipzig 1818–1889, vol. 2, pp. 135–137.

[2] [Cf. Michael Stolberg, "Emotions and the body in early modern medicine", in: *Emotion review* 11 (2019), pp. 113–122; in the original, German publication of this

M. Stolberg, *Medicine and the Body in Early Modern Europe*, Palgrave Studies in Medieval and Early Modern Medicine,
https://doi.org/10.1007/978-3-031-84556-7_2

diseases. Physiological and pathological theories had to interpret affects convincingly if they wanted to prove their worth. But above all, affects were thought to play a key role in the genesis of numerous illnesses. Physicians and laypeople alike counted them among the most important causes of disease,[3] including not least of all those illnesses like epilepsy and convulsions which many contemporaries might interpret as having been "cast" upon the patient or as the result of demonic possession.[4] This meant that medical dietetics had to pay close attention to the affects. In the effort to promote a healthy way of life so as to ward off disease or keep a present disease from worsening, the defense against the harmful influence of affects took an important position. Along with eating and drinking, sleeping and waking, exertion and leisure, air, and excretions, the affects counted as one of the six *res non-naturales,* which, according to a long tradition, had a decisive influence on disease and health.

Despite their great significance in science and everyday experience, the early modern perception and interpretation of affects has been given little attention in medical-historical research. While we are relatively well-informed about the theories of affects advanced by prominent philosophers, such as Descartes,[5] the predominant scholarly views concerning the physical manifestation and pathological effects of affects and the meaning of affects in everyday medical practice have so far been studied cursorily only.[6] When it comes to the subjective experience of medical laypeople,

article a forthcoming contribution to a collective volume was quoted which unfortunately never made it into print].

[3] This is a very old notion; see, in particular, Galen, *Opera omnia*, ed. by C. G. Kühn, Leipzig 1822 (repr. Hildesheim 1964), vol. 8, pp. 301–302 (*De locis affectis*, book 5, ch. 1) and ibid., vol. 5, pp. 360–427 (*De Hippocratis et Platonis placitis*, book 4).

[4] Petrus Burmannus (praes.), *De morbis passionum animi*. Subj. Christophorus Lipstorp, Leiden 1719, p. 6.

[5] Francesca Bonicalzi, *Passioni della scienza. Descartes et la nascita della psicologia*, Milan 1990; Susan James, *Passion and action. The emotions in seventeenth-century philosophy*, Oxford 1997; John J. Blom (ed.), *Descartes. His moral philosophy and psychology*, New York 1978.

[6] The few extant works focus on the post-Cartesian era: Lelland Joseph Rather, "Old and new views of the emotions and bodily changes. Wright and Harvey versus Descartes, James and Cannon", in: *Clio medica* 1 (1965), pp. 1–25; Walther Riese, *La théorie des passions à la lumière de la pensée médicale du XVIIe siècle*, Basel and New York 1965; Horst Bernhard Hohn, *"De affectibus animi" 1650. Die Affektlehre des Arztes Henricus Regius (1598–1679) und sein Verhältnis zu zeitgenössischen Philosophen*, Cologne 1990. The best overviews of the medical interpretation of the affects in the sixteenth

we are still scarcely able to guess what role this learned doctrine of affects had, how they experienced affects in times of both health and sickness, the fears they associated with them, or their attempts allay the possible dangers of affects.

With this in mind, the following article will seek to provide an overview of the period from about 1550 to 1700—albeit one that is brief and doubtlessly preliminary in some respects—of the learned discussion of affects and their role in the genesis of illness. In a first step, I will be drawing predominantly on anthropological, physiological, and dietetic treatises, and on physicians' *consilia* and case histories. In a second step, I will be turning to lay experience and lay interpretations of affects. For this, my analysis is supported by an examination of hundreds of early modern consultation letters penned by laypeople as well as approximately seventy early modern autobiographies written mainly in German.[7] Finally, I will be considering the commonalities and differences between the conceptions of physicians and those of laypeople.

The Physiology of the Affects

Within early modern learned literature, extensive discussion of the affects is mainly found in two quite different discursive contexts: first, in the general "natural philosophical", "anthropological"[8] accounts of the soul or the nature of man and the faculties of his parts and, second, in the extensive dietetic literature, where more consideration is given to the possible effects of emotions. These expositions are complemented in

and early seventeenth centuries still are Max Steinitzer, *Die menschlichen und tierischen Gemütsbewegungen als Gegenstand der Wissenschaft. Ein Beitrag zur Geschichte des neueren Geisteslebens*, Munich 1889, and H. M. Gardiner, Ruth Clark Metcalf, and John G. Beebe-Center, *Feeling and emotion. A history of theories*, Westport, CT, 1970 (first publ. 1937); some useful information can also be found in Anthony Levi, *French moralists. The theory of passions 1585 to 1649*, London 1964. Medieval notions of the affects have been described by Nancy G. Siraisi, *Taddeo Alderotti and his pupils. Two generations of Italian medical learning*, Princeton 1981, pp. 204–206 and pp. 226–229.

[7] I am very grateful to Brigitte Berger, Munich, who perused a large part of these autobiographies for relevant passages.

[8] Well-known examples are Otto Casmann, *Psychologia anthropologica, sive animae humanae doctrina*, Hannover 1594; Johann Sperling, *Anthropologia physica*, Wittenberg 1656; Albertus Kyperus, *Anthropologia*, Leiden sine anno, pp. 633–634.

manifold ways by the description and occasionally explanation of particularly striking cases as provided in the quite widespread *observationes* and *consilia* literature.[9] The more strictly speaking natural-philosophical, "anthropological" discussion of affects was controversial in some respects. In particular, the way the body and soul worked together, and the exact relationship between the body and the soul when affective phenomena were in evidence were matters of dispute. Being responses to external objects, the affects as a rule emerged from sensory perception and required the contribution of the soul. The soul alone could evaluate the sensory impressions, that is recognize them as pleasant or unpleasant, which was a prerequisite for a correlating affective reaction.[10] However, the soul was at the same time inextricably bound to the body during a person's lifetime, as physicians and natural philosophers never tired of pointing out.[11] It was not only *forma informans,* in an Aristotelian sense, but at least from a medical point of view, it was also *forma efficiens.* With the help of the vital spirits and animal spirits, as its direct instruments, the soul governed the body and played a key role as the ultimate director of all processes within the body.[12] This central "physiological" role of the soul and the differentiation of its various faculties with their respective seats in the body did not entirely dissolve the dichotomy between body and soul, which, contrary to a widespread misunderstanding, did not first appear with Cartesianism. The "physiological" account did, however, give this dichotomy a whole new meaning and it lessened the severity that it would come to have later. Even authors who conceded that the soul played a relatively dominant role in affective phenomena also continually emphasized the substantial contribution of physical changes. For example,

[9] On the genre, see Pedro Laín Entralgo, *La historia clinica. Historia y teoria del relato patografico*, Barcelona 1961.

[10] According to Vives, a proper judgment was not necessary, however; a movement of the power of imagination that was similar to a judgment was often sufficient (Juan Luis Vives, *De anima et vita libri tres*, Basel 1543, p. 208).

[11] Cf. Pierre Charron, *De la sagesse. Trois livres*, Paris 1604, p. 111: "L'ame qui n'est qu'une au corps"; Lemnius somewhat more cautiously argued that body and soul were "sympathia quadam ac consensu inter se connexa" (Levinus Lemnius, *De habitu et constitutione corporis, quam Graeci krasin, triviales complexionem vocant libri duo*, Jena 1587, preface).

[12] Cf. Katharine Park, "The organic soul", in: *The Cambridge history of Renaissance philosophy*, ed. by Quentin Skinner, Eckhard Keßler, and Jill Kraye, Cambridge 1988, pp. 464–484.

A. Luisinus defined affect as a "movement of the soul that does not listen to reason", simultaneously stressing the close connection ("copula") between the soul and the body.[13] F. N. Coeffeteau described "passion" as a "movement of the sensory appetites, caused by the perception or imagination of something good or bad" which is followed by a change "that happens to the body against the laws of nature".[14] On the other hand, the interpretation that is ascribed to the Stoics of the affects as mere incorporeal *opiniones*—was usually rejected as unacceptable. In Melanchthon's words: "Opinion is in the brain, affect in the heart".[15] Even Georg Ernst Stahl, who decidedly underscored the primacy of the soul, conceded around 1700 that "the different dispositions of the body and its humors can often be the cause of different affects". He did add, however, that "the physical constitution [usually follows] the characteristic features of the soul".[16] Some authors also advocated an understanding of the affects that is rather broad by today's standards, which left space even to the positive or negative experience of their bodily state as such. For example, joyfulness according to Cornelis Bontekoe promoted not only physical health but was also for its part sometimes owed to physical wellbeing: thus, one was filled with joy when one's body was healthy, one's appetite strong and one's muscles lithe, and when one was not easily overwhelmed by fatigue and sluggishness. Conversely, sadness ("tristitia") stemmed

[13] Aloysius Luisinus, *De compescendis affectibus, per moralem philosophiam et medendi artem tractatus*, Basel 1562, p. 14.

[14] F. Nicolas Coeffeteau, *Tableau des passions humaines. De leurs causes, & de leurs effets*, Paris 1631, p. 2; the same definition is found almost verbatim in Jean-François Senault, *De l'usage des passions*, Leiden 1658, p. 18, and in Samuel Antoni, *Medicina affectuum*. Praes. Joh. Andreas Schmidt, Jena 1694, pp. 4–5; for a list of ancient definitions see Johannes Vorstius, *Exercitatio physica de appetitu sensitivo et affectibus*. Proponit Hartmannus Koch, Rostock 1651; for on overview of the place of the affects in the history of philosophy and especially in Antiquity, see Silvia Vegetti Finzi (ed.), *Storia delle passioni*, Rome and Bari 1995.

[15] Philipp Melanchthon, *Liber de anima*, Wittenberg 1556, fol. 123r; Matthaeus Dresserus, *De partibus corporis humani, et de anima, eiusque potentijs libri duo*, Wittenberg 1581, p. 70.

[16] Georg Ernst Stahl, *Über den mannigfaltigen Einfluß von Gemütsbewegungen auf den menschlichen Körper*, Halle 1695. Translation and commentary by Bernhard Josef Gottlieb, Leipzig 1961, pp. 24–38.

primarily from illness coming from "that inner sensation which attacks the spirit if the body suffers even the least little bit".[17]

Likewise pointing to the body in the final assessment was the fact that, as experience taught, different people reacted very differently to the same object or the same situation, and this could only be explained with reference to different individual bodily constitutions. After all, individual characteristics, including character traits, were only to a very limited extent connected to the "soul" in the conception of the day—quite the opposite of today's understanding of the "psychological". The human rational soul was unchanging and immortal and indeed, according to some Protestant theologians and philosophers, all men ultimately had the same soul which originated with Adam and was simply passed on from generation to generation.[18] For this reason, what we understand today as "individuality" could then be primarily explained—if the term can be sensibly applied here at all—in terms of the different physical constitutions, the temperament, the diverse experiences, injuries and illnesses that left their mark on the body over the course of a person's life, the power of nature or the "inner heat", later also the individual properties and the faculties of the organs and fibers.[19] We even find repeated citation of the radical Galenic dictum according to which the "habits of the mind" follow the temperament of the body.[20]

[17] Cornelis Bontekoe, *Tractatus ethico-physicus de animi et corporis passionibus earundemque certissimis remediis*, ed. by Johannes Flenderus (= appendix to Arnold Geulincx, Gnothi seauton sive ethica), Amsterdam 1696, pp. 14–15.

[18] On traducianism and Lutheran anthropology see Markus Friedrich, "Das Verhältnis von Leib und Seele als theologisch-philosophisches Grenzproblem vor Descartes. Lutherische Einwände gegen eine dualistische Anthropologie", in: *Spätrenaissance-Philosophie in Deutschland 1550–1650*, ed. by Martin Mulsow, Tübingen 2009, pp. 211–249.

[19] In the course of the early modern period, the notion of "temperament" more and more moved away from a narrow understanding as the literally idiosyncratic mixture of humors and primary qualities (cf. Michael Stolberg, Homo patiens. Krankheits- und Körpererfahrung in der Frühen Neuzeit, Cologne and Weimar 2003, pp. 116–121); Georg Ernst Stahl, (Stahl, Über den mannigfaltigen Einfluss [see note 16], p. 26), for example, defined "temperament" as the "mutual relationship between the solid and the fluid parts".

[20] "Animi mores corporis temperaturam sequuntur". For example, in Thomas Wright, *The passions of the mind in general*. Ed. by William Webster Newbold, New York and London 1986, p. 117; cf. Martino Menghi and Mario Vegetti (eds), *Galeno. Le passioni e gli errori dell'anima. Opere morali*, Venice 1984.

Even the influential theory of affects developed by René Descartes, who is often (mis)understood as the founder of modern mind–body dualism, while it advertises a fundamentally different understanding of the soul, in no way negates the role of the physical in affective phenomena. Quite the contrary: at the heart of his theory of affects is the assumption of highly complex interrelations, which goes so far as to posit a primarily physical origin of certain affects. According to his account, the sensory impressions, coming from an encounter with a pleasant or unpleasant object, moved the animal spirits in the nerves and cerebral ventricles which, for their part, moved the pineal gland and in this way produced joy, shame, anger, and so forth in the soul. The same movements of the animal spirits and thus the same affects of the soul could, however, stem from changes within the body, for example when good, healthy blood spreads out in the heart and relaxed the nerves in a pleasant manner. This caused the animal spirits to move in the same way, and, via the pineal gland, this ultimately led to joy in the soul. Finally, the mere act of imagining joy moved and changed the heart and the remaining body parts through the animal spirits; the sensory impressions produced in this way then likewise generated the affect of joy through the pineal gland, as sketched out above.[21] The post-Cartesian theory of affects accordingly was often characterized by a particular attention to the animal spirits and by the repercussions of physical changes for the perception or experience of affects, as facilitated by the animal spirits. Thus, the theologian Gummarus Huygens defined the "passio animae" in a narrow, philosophical sense as an "affect of the soul that derives directly from a hidden physical movement of the spirits, the blood, the constriction and dilation of the heart and so forth within the body", and for further explanation he pointed to the joy that resulted simply from the healthy condition of the body.[22]

[21] René Descartes, *Les passions de l'âme*, Amsterdam 1649; I am drawing here, in particular, on Susan James' analysis (James, Passion [see note 5], pp. 94–108); in the wake of Descartes, Henricus Regius defined the "affectus animi" as an act of thinking ("cogitatio") that was linked to a vehement movement of the animal spirits in the cerebral ventricles, which strongly affected mind and body. Via the nerves, the resulting movement of the heart and the other parts of the body had a direct effect, in turn, on the brain and the mind (Henricus Regius, *De affectibus animi dissertatio, Utrecht 1650*. Repr. and transl. in Hohn, De affectibus [see note 6), pp. 199–235, here p. 205).

[22] Gummarus Huygens, *Breves observationes de actibus humanis et passionibus animae*, Liège 1694, p. 243; see also Cornelius ab Hogelande, *Cogitationes, quibus Dei existentia,*

Affects as the Cause of Illness

Regardless of how exactly one conceived of the processes and interactions, one thing was beyond question: as everyday experience taught, affects were often accompanied—if not characterized—by relatively pronounced bodily changes. Grief consumed people, anger reddened the face, fear and fright made one go pale. This also meant that no matter how much an independent, perhaps even dominant role was ascribed to the soul, the affects were most certainly linked to strong movements or changes in the body. And thus, crucial to our argument, they were able to influence very directly and powerfully the natural processes and functions within the body and to cause disease.

The connection between affect and disease already found expression in the terminology. The terms "affectus" as well as "passions" and the Greek equivalent "pathemata" traditionally denoted individual pathological states in a very general sense.[23] Terms such as "affectus hypochondriacus" were widespread in the eighteenth century, as was the German term "Pathologie", which served as the opposite of "Physiologie", but was also occasionally—and tellingly—used to refer to the doctrine of the *affectus animae*.[24] Also, the characterization of the affects as "accidents", that is *accidentia*, points to a central concept in the pathology of the period,[25] and some authors preferred from the outset the term "perturbationes animae",[26] or even defined the affects as a "status praeternaturalis sive morbus", as "diseases by which it [the soul] is plagued and thrown into disarray".[27] This is to say that the medical

item animae spiritalitas et possibilis cum corpore unio demonstrantur, Leiden 1676, pp. 98–99.

23 For this reason, the term was often specified, in our context, by adding "animi" or "animae".

24 Maturinus Castanaeus (praes.), *Pathologia generalis. Seu dissertatio in genere de affectibus animae*. Def. Daniel Mathesius, Copenhagen 1692; it is not clear, whether the *praeses* Castanaeus or his student Mathesius authored this work.

25 Ambroise Paré, *Les Œuvres*, 9th edn., Lyon 1633, p. 29.

26 Johannes Hadrianus Slevogt, *Diatribe physiologica de affectibus animi tribus dissertationibus medicis singularem irae, metus atque tristitiae utilitatem per rationes et experientiam demonstrantibus praemissa*. Exponit Jo. Georgius Möllerus, Jena 1695, p. 7.

27 Coeffeteau, Tableau [see note 14), p. 17; the original French reads "les maladies dont elle est trauaillee et troublee".

discussion of the *passiones* was much closer in meaning to disease than it was to Aristotle's twin concepts of *actio* and *passio*.[28]

During the period in question, an abundance of diseases was associated with the affects. In the *observationes* and *consilia* literature of the early modern period, physicians described particularly striking cases and were fond of citing them as evidence in their theoretical treatises.[29] Some authors collated numerous diseases that resulted from certain affects in their experience.[30] For example, Peter Burmann, listing illnesses that resulted from grief, cited (among others) colics, cholera, jaundice, hypochondria, melancholy, scirrhe (indurated tumors), obstructed menstruation and lochia, dropsy, swollen feet, general weakness, constipation alternating with loose stools, consumptive and hepatic fever, as well as chronic diseases in general.[31] In the same place, he provided an early description or at least an intimation of that state which we today refer to as "hospitalism". He wrote that infants and children who were left in the care of pilgrims to whom they could not become accustomed became increasingly weak and emacatied, due to their grief ("moeror"), and stopped growing. When they were returned to their parents, however, they began to thrive again.[32]

The affects were not considered harmful per se. On the contrary, numerous authors objected resolutely to a general condemnation of the affects, a position that was attributed to the Stoics above all.[33] For

[28] Cf. James, Passion [see note 5); Coeffeteau, Tableau [see note 14), pp. 17–18, explicitly pointed out the difference. In this context, "passion" did not mean that a subject acquired a new form or a new quality but referred to a change, which ran counter the "constitution et disposition naturelle".

[29] Authors commonly limited their account to the pathological effects of the most important, most common, or most intense affects and, in partcular, to joy, love, fear, wrath, and sadness. Shame, envy, and jealousy, leave alone gratitude, humulitiy, scorn, pity, or adoration often were not even mentioned or, at best, described as similar to one of the basic affects.

[30] In this sense, Giorgio Baglivi called for a systematic collective endeavor to find out, which affects primarily caused which diseases ("welche Kranckheiten vornemlich von einer ieden Gemüths-Bewegung entstehen"); see Giorgio Baglivi, *Zwey Bücher De Praxi medica, wie solche in klugen Observiren wiederum auf den alten Fuß unserer fleißigen Vorfahren möge gestellet werden*, Lübeck and Frankfurt 1705, pp. 251–252).

[31] Burmannus, De morbis [see note 4), p. 31.

[32] Ibid.

[33] Thus, for example, emphatically Bontekoe, Tractatus [see note 17), pp. 37–39.

example, joy, hope, and a clear conscience—as Matthaeus Dresserus put it in 1581—were a sweet sustenance for health and life; in fact, without joy, hope, and love, there was no life.[34] Rather, what was crucial for a medical assessment of the affects was their intensity and relation to the individual physical constitution. A dangerous situation arose if a person's affects were excessive and intense, especially if they exacerbated existing pathological tendencies, for example when cholerics became "heated".

Furthermore, the "negative" affects, such as anger, envy, or grief, were thought to be far more dangerous in general than "positive" affects such as joy and love. For this reason, they dominated by far in medical case histories. Yet, even a strong "negative" affect could occasionally have a curative influence. "Cold" natures, for example, might be agreeably warmed by a hot surge of anger, while "hot" natures were soothingly cooled by grief.[35] Conversely—and here, too, physicians were able to provide examples—a sudden positive affect such as excessive joy could have fatal consequences. For this reason, Valles advised that one should give the mother of a son who was believed dead some sad news before giving her the joyful news of her son's unexpected survival, to keep her from dropping dead.[36]

Often, the medical ideas regarding the exact processes by which affects exerted a harmful influence in the body were only loosely adumbrated. In many cases, authors were content to point simply to the common, generally recognized experience that strong affects, particularly negative ones, could have disastrous consequences. However, from the plethora of descriptions, and from several quite detailed treatises

[34] Dresserus [see note 15), pp. 69–70; quoting Descartes, Slevogt, Diatribe [see note Anm. 26], p. 3) wrote in the late seventeenth century, "passiones animae ad salutem et conservationem mortalium a natura concessae sunt, ut corpus ad rerum utilium prosecutionem et noxiorum fugam comparetur".

[35] Johan van Beverwijck, *Schat der Gesontheyt*, Amsterdam 1672, p. 26; fear of death was said to sometimes work virtual miracles. François Valleriola told the story of Giovanni Berla, for example, who had been paralyzed and bed-ridden for years and was considered a desperate case, after many treatment attempts had failed. One day, a fire broke out, however, and in his fright, the patient ran from the house and from this day on, sensation and mobility returned into the affected parts (François Valleriola, *Observationum medicinalium libri sex*, Lyon 1573, pp. 77–83).

[36] Franciscus Vallesius, *De sacra philosophia, siue de iis quae physice scripta sunt in libris sacris*, Frankfurt 1600, p. 426.

with a stronger natural-philosophical orientation, a relatively differentiated picture emerges. Playing an essential role in the understanding of the "physiology" of rapid, sudden affects and their potentially pathogenic consequences was the traditional doctrine of the vital and animal spirits, as well as the closely related doctrine of the vital, "implanted heat" with its material substrate of vital moisture, the *humidum radicale*. These explanatory elements were often used in an eclectic manner, however, and were sometimes inconsistent with other concepts, such as the doctrine of qualities, but also with specific theories about the genesis of particular diseases. One at times gets the impression that physicians were dipping into the fund of inherited explanatory approaches as if it were a kind of quarry from which they could take at random that which seemed to best suit their particular case. Therefore, one searches in vain for a comprehensive or even generally recognized medical theory of affects. All that can be identified are certain widespread explanatory approaches, terms, and images.

In the sixteenth and early seventeenth centuries, sudden, quick affects were primarily seen in a very literal way as accompanied and characterized by movements, in particular by movements of the heart, the traditional seat of the *anima sensitiva* and the *irascible* and *concupiscible* faculties, and of the vital spirits and humors that flowed from the heart to the periphery. Here, two basic types of affects could be distinguished. Some affects—the most important examples being joy and love—were defined or accompanied by a movement of the humors and vital spirits from the heart out toward the periphery. Externally, this manifested, for example, in the skin becoming hot or in blushing; the pulse, expressing the *spiritus* movement, became faster. As opposed to this, there were those affects which were defined or accompanied by the *spiritus* and/or humors retreating to the heart from the rest of the body. Thus, if a person got a fright or felt fear or grief, the extremities became very cold and white, while the humors and spirits pushing into the heart elicited feelings of pressure or tightness. The Nuremberg physician Fabricius wrote in this vein that fear drove "the spirits back into the body with a quick fervor, which soon causes one to grow pale in the face, to have cold fingertips, to start trembling in every limb, to sometimes develop diarrhea, to falter in one's speech and feel heart palpitations".[37]

[37] Johann Georg Fabricius, *Consilium diaeteticum* [1658], Universitätsbibliothek Erlangen, Ms. 948/II, foll. 435r-463r, here fol. 459v, quoting Ambroise Paré.

In this way, the dangers of "negative" affects found a physiological explanation in the movement of the humors and spirits which rushed away from the unpleasant object into the interior: according to the old Galenic conception, the humors and spirits by nature moved only in one direction, from the heart and liver to the other body parts, which they continually nourished and warmed. Thus, fright, fear, or grief produced a movement that ran counter to the natural movement and distribution of the humors and *spiritus* in the body, disrupting them severely. At the same time, the heart might fill to such an extent that it could no longer carry out its duties. The blood and the spiritus "suffocated" the heart.[38] Much more rarely, positive affects reinforced the natural movement away from the heart to such a degree that the heart emptied all too much. This explained the fatal consequences of excessive joy. As Fabricius presented the account put forth by the famous Jean Fernel, if "great and unusual joy chases the subtle vital spirits along with the natural heat out of the heart and into the outer parts of the body", then the heart is robbed all too much of the humors and above all of the life-sustaining spirits.[39]

However, a movement of the blood and vital spirits away from or toward the heart explained some affects better than others. Shame, for example, did not fit in well with the explanatory framework sketched out above. It arose from an encounter with a negative object (or an unpleasant situation) and was therefore similar to fear.[40] The shameful reddening of the face, however, suggested that the humors and spirits were flowing away from the heart. Indeed, Levinus Lemnius put shame on par with anger and joy and explained sudden death from intense humiliation as the consequence of a dramatic emptying of the heart of blood and vital spirits.[41] Ambroise Paré postulated a two-part movement: first, a brief movement into the heart and then a movement away from the heart into the periphery, which reddened the face.[42] The issue of anger raised similar questions. Some people turned pale with anger; however, the typical redness in the face pointed to a displacement to the periphery—in other words, to a movement toward the negative object rather than away from

[38] Lemnius, De habitu [see note 11], fol. 64r.

[39] Fabricius, Consilium [see note 37], fol. 461v.

[40] See, for example, Coeffeteau, Tableau [see note 14], p. 21.

[41] Lemnius, De habitu [see note 11], fol. 63v.

[42] Paré, Œuvres [see note 25], p. 30.

it. Indeed Magirus, for example, held that the vital spirits pressed toward the negative object in the case of anger—so as to chase it away.[43]

In any case, the pathological impact of long-lasting affects—affects that endured over the course of months or even years—could not be adequately explained by a sudden change in the movement of the humors and vital spirits.[44] Here, authors liked to take recourse to a different explanatory tradition, that is the ancient metaphor of "inner heat" as an oil lamp. Vital heat was described as a kind of vital flame which in the course of a person's life increasingly consumed its fuel, the *humidum radicale,* and ultimately went out, just as an oil lamp gradually uses up its oil. This was the common explanation for the aging process and natural death at the time. The vital flame could furthermore be suffocated by an excess of humors or, as with fever[45] for example, could be overpowered by a "foreign heat". Physicians were very familiar with this imagery, and the typical symptoms that accompanied grief and sorrow in particular were proof of the connections here: as experience taught, sad people afflicted by grief were recognizable by their gaunt, careworn appearance—they were literally wasting away with sorrow. Grief accelerated the natural aging process and made those afflicted with it wither and die before their time.[46] The sudden graying of the hair, as was sometimes described, was only one particularly striking change. Sorrow and grief slowly consumed the body's substance, caused it to become dry and cool down, robbed it of the moisture and vital heat that was necessary for continued life. It was with good reason, then, that the poets, according to the Nuremberg physician Johann Georg Fabricius, spoke of "eating, devouring, pilfering, consuming" worry.[47] On closer examination, course, tensions, and fissures can be discovered here as well, however. After all, a life-shortening effect should have been the result of those affects in particular that, like anger, accelerated the consumption of the *humidum radicale* by making the vital flame

[43] Johannes Magirus, *Physica peripatetica ex Aristotele, eiusque interpretibus collecta*, Frankfurt 1593, p. 344.

[44] In this sense, Magirus, Physica [see note 43], p. 343, attributed the slow but in the long run no less harmful effects of sadness to the constant slow movement of warmth inside the body.

[45] Avicenna, *Canon medicinae*, Venice 1556, book 4.

[46] Vallesius, De sacra philosophia [see note 36], pp. 425–427; Bontekoe, Tractatus [see note 17], p. 19.

[47] Fabricius, Consilium [see note 37], fol. 460v.

of the "implanted" inner heat burn stronger and stronger, so to speak. Indeed, Fabricius explained the harmful consequences of anger by referring to yellow, hot bile that resulted from it and penetrated "different places in the body" and heated "the blood and vital spirits". Through this, anger ignited a fever and drove "the natural heat out, into the limbs", harming in particular sufferers of gout, as was described in the well-known little verse, "Zorn, Wein vnd Weiber verderben unser Leiber" ("Wrath, women, and wine, throw our bodies to the swine"). When a person got angry, this also ignited "the subtle little spirits, and they are consumed by it". For this reason, he advised avoiding anger "like a poisonous beast that wants to strangle people".[48] And so, some authors placed more or less unconnected explanatory elements alongside one another, which together were to explain the influence of particular affects. According to Paré, grief dried the body out, but also made the spirits cloudy, constricted the heart so that it could not produce sufficient spirits and the few it did produce it could not distribute adequately in the body; as a result, the vital faculties became weakened.[49]

Regardless of the inconsistencies and contradictions we encounter here, one aspect of the medical theories of affects does come to the fore very clearly: the heart was the crucial seat of affective phenomena and moreover of their pathogenic effects. The Old Testament already confirmed this. Citing from the Book of Proverbs, Valles wrote that just as a worm damaged wood, grief damaged the hearts of men.[50] While the senses and the brain were crucially involved in affective phenomena, they, too, ultimately took effect through the heart. According to Thomas Wright, for example, the animal spirits, reacting to a pleasant or unpleasant object, flowed from the brain through certain secret ducts to the heart. There, they announced to the heart the presence of the object in question, which elicited an immediate reaction, causing the heart to either pursue or avoid the object.[51] This close relationship to the heart also made the affects dangerous. After all, the heart was considered the decisive seat of the vital faculties, as the place of that "innate" inner heat from which, in Galenic physiology, the warm vital spirits flowed into

[48] Ibid., fol. 458r.

[49] Paré, Œuvres [see note 25], p. 30.

[50] Vallesius [see note 36], p. 425.

[51] Wright, Passions [see note 20], p. 123.

the rest of the body and vitalized it. Accordingly, early modern physicians tended to administer "cardiotonic" remedies, so-called "confortativa cordis", in cases where an illness was thought to be caused by affects.

The affects were thus also in a very literal sense "movements of the heart" or "motus cordis", as it was put sometimes,[52] and according to old Galenic doctrine the expansion of the heart chamber was also an active movement.[53] The term "emotions"—a term we find occasionally used in the seventeenth century—like the later German term *Gefühlsbewegungen* (movements of sentiment) was therefore not metaphoric but was originally understood in a very literal sense as a reference to the local movement of the heart or the humors or spirits. In other words, it denoted the result or physical manifestation of affects. In the contemporary use of the term "emotion", this aspect of movement manifests itself clearly. In 1633, Ambroise Paré for example warned that even the surgeon must take care not to underestimate the "accidents" of the soul ("accidens de l'âme"). After all, they caused "de grandes émotions". Joy, hope, and love widened the heart and moved the blood and the spirits, and the smiling face became rosy and joyful.[54] Similarly, Senault in 1658 underscored that the passions went against the laws of nature because they attacked the heart, something that inevitably became visible in the "esmotion" of the other body parts.[55] And Charron called for an analysis of the affects with a view to their "esmotion et impetuosité".[56]

[52] Magirus [see note 43], p. 342; in similar terms, Casmann [see note 8], p. 406, defined the affects as "cordis motiones ab animali facultate motrice procedentes, quibus cor interna sanguinis et spirituum commotione, cognitionem et appetitionem sensitivam sequente, afflicitur."

[53] Rudolph E. Siegel, *Galen's system of physiology and medicine*, Basel and New York 1968, pp. 31–32.

[54] Paré, Œuvres [see note 25], p. 29; see also Coeffeteau, Tableau [see note 14], p. 15: "tout le corps se sent esmeu".

[55] Senault, De l'usage [see note 14], p. 20. A related term, which could also refer to the soul, however, was "commotion". In this sense, Isaac Watts, in the early eighteenth century, called the passions "commotions of animal nature", which affected the heart, in particular (Isaac Watts, *Discourses of the love of God and the use and abuse of the passions in religion*, London 1729, p. 9).

[56] Charron, De la sagesse [see note 11], p. 111; in another passage (ibid., p. 556) he also used the term in the older (and then still very common) sense of "turmoil"; this understanding of the term sometimes still emerges in the early nineteenth century. William Hall, for example, distinguished the "passions" from the "emotions" as the "sensible effects", "which are perceived to arise from the predominance of a particular passion

As Harvey's model of blood circulation gradually gained recognition in the seventeenth century, the interpretation of the affects as movements of the heart and/or the humors and the spiritus did not simply become obsolete.[57] However, because, compared to Galenic physiology, the movement of the blood was taken to be much more rapid and circular, additional explanatory elements were required. The authors of works about affects proved to be quite inventive here. In Regius's 1650 depiction of joy, for example, the spirits of the soul flowed amply into the "widening" and "expulsive" fibers of the heart and its vessels, thereby causing an increased flow of warm blood in the body, accompanied by a reddening of the skin and a pleasant sensation of warmth.[58] Peter Burmann postulated that, in addition to the natural blood circulation, there was a slower *motus tonicus* of the vessels. As the blood circulated in a sustained manner, this *motus tonicus,* under the sway of the affects, distributed the blood in different ways in the body depending on the affect in question. In the case of grief, for example, it amassed blood in the heart, and the extremities became cold and pale.[59] Cornelis Bontekoe explained the pathological consequences of fear with reference to a predominantly contractive rather than a relaxing movement of the heart; because the blood would continue to come in from the lungs, the blood and humors flowed more slowly and thickened. It was for this reason that sad people hit their chests with their hands—to stimulate the movement of the stagnating blood.[60] Similarly, Leopold Neystifter, taking

or sensation"; as an example, he pointed at the effects of fear on the face and at the movements the coward made facing danger (William Hall, *Dissertation on the morbid effects induced on the mind and body by grief & fear*, Philadelphia 1812). According to Kingsmill Davan, some authors used the term "emotions", at the time, "to express the corporeal feelings about the praecordia, which are termed feelings of the heart" (Kingsmill Davan, *An essay on the passions*, London 1799, p. 105).

[57] Underestimating the continuing influence of the older notions and their integration into the new physiciology, Barbara Duden has argued that Harvey's new physiology made the sensible qualities that used to be ascribed to the affects as movements of the heart obsolete and turned them "into a pure metaphor" (Barbara Duden, "Anmerkungen zur Kulturgeschichte des Herzens", in: *Von der Auffälligkeit des Leibes*, ed. by Farideh Akashe-Böhme, Frankfurt am Main 1995, pp. 130–145, cit. p. 136).

[58] Regius, De affectibus [see note 21], p. 212.

[59] Burmannus, De morbis [see note 4], p. 16 and p. 31.

[60] Bontekoe, Tractatus [see note 17], pp. 20–21; he attributed this increased contractive movement of the heart to a redistribution of the spirits in the body.

an anti-Cartesian standpoint, primarily located the effects produced by affects in a changed contraction of the heart's fibers and vessels. In the case of hatred, the fibers contracted and hindered the flow of the blood and humors, while with joy, they widened.[61]

To explain the unique and especially powerful impact of affects in particular cases of illness, physicians occasionally presented theories that went even further. Nicolaus Tulpius, for example, saw fits of anger, especially as experienced by women in childbed and old people, as an important cause of hemiplegia and paraplegia. Their brains being particularly weak, a humor could easily dislodge there, and it would not so much be dissolved by the rising bile, the "immediate cause of anger", as driven into the base of the spinal cord. The spinal cord would then be obstructed, either unilaterally or completely—the implication being that this prevented the passage of the animal spirits, which controlled the limbs.[62] For his part, Magirus explained that the almost senseless fury of the irascible was the result of bilious humors which, due to their heat and lightness, easily entered the brain and robbed the affected person of his or her reason.[63]

We see how a whole host of different explanatory concepts was brought to bear in the early modern discussion of affects and their influences. Few areas of the contemporary medical discourse were characterized by such pronounced pluralism.[64] The doctrine of temperaments, however, based as it was in humoral pathology—and this may surprise some readers—was of secondary importance or even no importance at all when it came to understanding affective phenomena and their pathological influence. This doctrine hardly offered any theoretical starting points anyway that would have explained sudden, rapid affective responses and

[61] Leopold Neystifter, *Theatrum passionum animae*, Innsbruck 1721, p. 166 and pp. 201–202.

[62] Nicolaus Tulpius, *Observationum medicarum libri tres*, Amsterdam 1641, pp. 130–132.

[63] Magirus, Physica [see note 43], p. 344; cf. Stolberg, Emotions [see note 2], on the elaborate explanations of the effects of the passions which Leonhart Thurneisser provided.

[64] In this context, the manifold "physiological" explanations of typical bodily manifestations of the different affects, such as tears and sighing, should be mentioned, which were not considered as pathological.

the concomitant "movements" of the heart.[65] Not even the frequently named connection between anger and yellow bile withstood close inspection because, as Timothy Bright dryly noted, the bodies of jaundiced people were positively flooded with yellow bile, but these men and women were certainly no angrier than anyone else.[66] The significance of the doctrine of the humors and temperaments lay primarily in the explanation of the individual inclination toward certain affects. In contemporary terminology, temperament was only a *causa remota*.[67] Not only this, but the doctrine of temperaments was in competition with other explanatory approaches, beginning with the oft-cited pseudo-Aristotelian claim that animals with small hearts were more courageous than those with large hearts because with them, the heat was more concentrated and the blood that flowed in was accordingly heated more quickly.[68]

There was fairly general agreement about one thing, however: cholerics—people whose bodies were ruled by hot and dry yellow bile—inclined toward anger, while melancholics—dominated as they were by cold and dry black bile—inclined toward sorrow. Joy, for its part, was primarily associated with warm and moist blood.[69] Because different humors and temperaments were thought to dominate depending on age, sex, and climate, individual affects could be attributed to different people, depending on their sex, age, and region. According to Slevogt, for example, French- and Englishmen with their sanguine temperament did not get particularly vexed if their wives engaged in conversation with a young man; Italian men, however, with their inclination toward black bile, responded with the greatest indignation.[70] A matter of particular dispute

[65] In this sense, Timothy Bright, *A treatise of melancholie*, London 1586 (repr. Amsterdam and New York 1969), p. 85, wrote: "these passions being wrought of the heart by a certaine enlarging of it selfe, if it be pleased, and closing, if it be contrarily affected: which be actions not of complexion, but of frame & shape". He explained the connections in more detail in his discussion of the effects of anger, which the medical tradition associated especially closely with yellow bile and an upwelling in the heart: "first the heart moueth, kindled with anger, then the bloud riseth, which being cholericke encreaseth the heate, but addeth nothin to the passion" (ibid., p. 88).

[66] Ibid., p. 88.

[67] Thus explicitly, Slevogt, Diatribe [see note 26], p. 15.

[68] Pseudo-Aristotle, *Problemata Aristotelis, ac philosophorum medicorumque complurium*, sine loco 1558.

[69] Coeffeteau, Tableau [see note 14], p. 23.

[70] Slevogt, Diatribe [see note 26], p. 15.

was the inclination toward different affects in the two sexes. This was an important subject of the "querelle des femmes", which since the fourteenth century had brought forth numerous works and found numerous physicians as participants, who mostly took the side of the "gynephiles". Thomas Wright, for example, wrote that women "by nature" inclined more toward pity and mercy than men because their delicate physical constitution ("complexion") inclined them more toward sympathy. Due to their deficient heat and their "native shamefastness", that is, inborn reserve, they also had, in his view, better self-control than men.[71] With respect to the genesis of illnesses borne of affects, however, all of this was of limited significance.

Affects and Illnesses in Lay Medical Culture

If we turn to the experiences, interpretations, and practices of early modern laypeople as we encounter them in patient letters and other ego-documents, we find first of all that there was a generally held conviction of the affects' paramount influence on health. At least among people from educated circles—those who were able to leave such written testimonies behind—illnesses (their own and those suffered by others) were described time and again as the result of strong affects. Conversely, some patients expressly emphasized that the illness in their case was not caused by an affect, that their office or household did not cause them any trouble and that therefore a different cause had to be assumed.[72] In particular, grieving a lost loved one was frequently seen as related to a subsequent severe illness and conversely, the intensity of the mourning was stressed by pointing to the severe illness that ensued. When Glückel of Hameln and her husband's two-year-old child fell ill, developed swollen hands and feet, and died four weeks later, following "great complaints and suffering", the couple grieved so much "that both of us endured severe illnesses for a long time; we had that from our great sorrow".[73] Ludwig von Diesbach gave a similar account about the family of the ten-year-old son of

[71] Wright, Passions [see note 20], p. 119.

[72] Staatsbibliothek Berlin, Ms. germ. fol. 423a, foll. 311r-314r, letter from Johan Lindt, Easter 1580.

[73] Glückel von Hameln, *Denkwürdigkeiten der Glückel von Hameln*. Transl., comment., and ed. by Alfred Feilchenfeld, Frankfurt am Main 1987, p. 108.

a cousin who died in an epidemic: "And the Rüssegg woman was pregnant and was to give birth soon. This honorable, upright woman felt so much sorrow that she became sick and died".[74] Theodor Fabricius even fell into such a deep state of grief in 1547 after the death of his wife that he wandered through the villages at night and feared he would have a stroke.[75] Sometimes the effects were felt in a very specific manner. Simon Roter, for example, reported how, when an unexpected death occurred, he "was dealt great grief and sorrow, so much so that this past 28th of November, around 1 o'clock at night, I had a paralysis in the left leg".[76]

Alongside grief, fear and fright were often named as a cause of illness. Gottschalk von Weinsberg traced his protracted fever to the fright he got when a spider suddenly fell into his drink.[77] Caspar Hirsch described how his wife gave birth to a dead child after a severe fright.[78] One young epileptic's medical history began with a "phrenesia" and a "syncope and loss of consciousness" which he suffered as a child when he was frightened "by a simpleminded person or fool". At school as a seven-year-old, he "again fell unconscious, frightened by an unexpected noise". He began to experience a sudden loss of speech while in the midst of talking. He suffered his first convulsive seizure at the age of eleven during mass.[79] And there were even two women who were said to have encountered a "werewolf" in Onolzheim in 1685 and to have died not because they had been injured—three other women actually suffered bites and survived—but from the terrible fright.[80]

[74] Ludwig von Diesbach, *Die autobiographischen Aufzeichnungen Ludwig von Diesbachs. Studien zur spätmittelalterlichen Selbstdarstellung im oberdeutschen und schweizerischen Raume*. Ed. by Urs Martin Zahnd, Bern 1986, p. 35.

[75] Theodor Fabricius, "Lebensbeschreibung des ersten anhaltischen Superintendenten". Ed. and transl. by Dr. Münnich, in: *Zerbster Jahrbuch* 16 (1931), pp. 37–94.

[76] Staatsbibliothek Berlin, Ms. germ. fol. 421a, foll. 368r-369r, letter from Roter, Sunday after Mary's conception, 1576.

[77] Herrmann von Weinsberg, *Das Buch Weinsberg. Aus dem Leben eines Kölner Ratsherrn*. Ed. by Johann Jakob Hässlin, Stuttgart 1961, p. 261.

[78] Caspar Hirsch, "Familienaufzeichnungen". Ed. by Ferdinand Mencik, in: *Jahrbuch der Gesellschaft für die Geschichte des Protestantismus in Österreich* 22 (1901), pp. 18–52, cit. editor's note on p. 22.

[79] Staatsbibliothek Berlin, Ms. germ. fol. 424, foll. 365r-367v, letter from Dr. jur. Folckhamer, May 24, 1582.

[80] Staatsarchiv Nürnberg, Ansbacher Oberamtsakten, final report, October 10, 1785.

The ravages of war gave rise to fear and fright in a particularly potent manner. Klara Staiger, writing about the first night of attacks when her town was conquered during the Thirty Years' War, described how everyone got such "a pitiful great fright" from the "shooting, screaming, and running that some fell ill over it and one died in childbirth, God comfort her dear soul".[81] Later, in 1632, her convent was plundered by soldiers. Her abbess received such a fright that it "went through all her limbs". The following night she complained that she felt cold. Soon she could scarcely breathe and then she fainted. The next morning, she came back to her senses, received her last rites, and passed away.[82]

Alongside fear and grief, professional set-backs sometimes caused or exacerbated illness for men. Perhaps this also reflects the comparatively higher value of "honor" in early modern society.[83] Christoph zu Dohna, who did not get along with the crown prince, believed he would find no more happiness in Brandenburg after the death of the father, the elector. "This thought", he wrote, "along with the keen feeling of pain over having lost my benefactor" oppressed him so much that he got jaundice or could have easily gotten jaundice, as he wrote, lessening the claim.[84] And it could even go further: according to the account of a contemporary of Hans Khevenhüller, the latter felt so aggrieved when the Emperor did not show him the kindness he had hoped for that two days later he became paralyzed and fainted and ultimately died.[85]

As concerns anger, Carol Stearn has observed that anger hardly finds articulation in early ego-documents of the early modern period and that it only gradually found increased acceptance in the course of the

[81] Klara Staiger, *Klara Staigers Tagebuch. Aufzeichnungen während des Dreißigjährigen Krieges im Kloster Mariastein bei Eichstätt*. Ed. by Ortrun Fina, Regensburg 1981, p. 313.

[82] Ibid., p. 41.

[83] Regius, De affectibus [see note 21], p. 230.

[84] Christoph zu Dohna, *Die Denkwürdigkeiten des Burggrafen und Grafen Christoph zu Dohna (1665–1733)*. Introd., comment., and transl. by Rudolf Grieser, Göttingen 1974, p. 71.

[85] "Relation des Georg Mayer, Buchhalter der Fugger, über den Tod Hans Khevenhüllers. Valladolid, 13. Mai 1606", in: Hans Khevenhüller, *Hans Khevenhüller kaiserlicher Botschafter bei Philipp II. Geheimes Tagebuch 1548–1605*. Ed. by Georg Khevenhüller-Metsch, Graz 1971, p. 312.

seventeenth century with the development of a new culture of individualistic self-assertion.[86] Her observation is essentially confirmed by German patient letters and autobiographies: while anger was always given due attention in the medical descriptions of the influence of affects, it hardly played a role in personal testimonies; at most, people occasionally recognized the harmful influence of anger in others. Henrich Brokes, for example, told of how anger became his father's undoing after he had hurt his shin on a piece of wood: "The injury was not dangerous. But because of several things that made him agitated three days later, he had a stroke, that is, he was paralyzed ["gerühret"] on one side".[87]

In the eighteenth century, there are occasional references to the effects of anger. One patient, for example, who was plagued by stomach and head complaints, wrote to Friedrich Hoffmann that he "inclined toward anger and zeal very much, which I feel immediately in my entire body".[88] A female patient wrote that it was her fault that she "can become angry easily and so bile may have entered the blood".[89]

Only rarely did laypeople make explicit their ideas about specific physical changes they experienced after strong affect, as in this account of bile entering the blood following anger. By contrast, it appears that the image of "consuming" grief was quite well-known. What "grief and gloominess do to man", wrote a clerical patient of Leonhard Thurneisser, "is shown in the divine writings", namely "how the marrow in the bones is consumed by it and people are robbed of all their natural strength, to which other complaints easily come added".[90] Usually, however, no more than the idea of an affinity between certain affects and certain diseases emerges from the sources. Apoplexy, for example, was widely associated

[86] Carol Z. Stearns, "'Lord help me walk humbly': Anger and sadness in England and America, 1570–1750", in: *Emotion and social change. Toward a new psychohistory*, ed. by Carol Z. Stearns and Peter N. Stearns, New York and London 1988, pp. 39–68.

[87] Henrich Brokes, "Aus dem Tagebuche des Lübeckischen Bürgermeisters Henrich Brokes". Ed. by Oberappellationsrat Dr. Pauli, in: *Zeitschrift des Vereins für Lübeckische Geschichte und Altertumskunde* 1 (1855), pp. 79–92, pp. 173–183, and pp. 281–347, and ibid. 2 (1858), pp. 1–37, pp. 254–296, and pp. 367–465, cit. p. 82.

[88] Friedrich Hoffmann, *Medicina consultatoria*, vol. 5, Halle 1726, pp. 300–301, signed "v. D.", March 10, 1727.

[89] Ibid., vol. 10, Halle 1733, pp. 279–282, letter from a 44-year old female patient (around 1732).

[90] Staatsbibliothek Berlin, Ms. germ. fol. 421a, foll. 193r-194v, letter from a court preacher (possibly Frid. Hartweg), Thursday after Bartholomew, 1576.

with an overfilling of the vessels, particularly in the head. In this light, it made sense that Brokes's father was experiencing apoplexy, given that his face was often red with rage. Bozenhart account of his encounter with a wolf was already comparatively detailed: he emphasized his great fear by saying that if someone had pricked him at that moment, not a single drop of blood would have issued forth; he was implying, it would seem, that his blood had retreated to his heart from fright.[91]

On the other hand, it is striking just how closely medical laypeople—not only in their belief in, at times fatal, physical effects of affects but also in the terms they used—connected affective experience with the heart as a physical, tangible body part. Today, when we associate the heart with emotions in manifold turns of phrase, we are mostly being metaphorical. In texts written by laypeople in the early modern period, however, the use of the heart in a metaphoric sense is often combined with descriptions of very real bodily sensations in and around the heart.[92] If a person's heart was described as "silent" or as "cheerful", this would presumably have been chiefly metaphorical then already. But when strong affects were connected with a "stab to the heart", or if the heart was simply about to "burst" from love, or if it was "all tightened up" upon saying goodbye, then the line between physical sensation and mere metaphor became obscured. Here, the affects could be experienced in the body as physical changes or movements of the heart. Not only this, but the same terms that from today's perspective would be read as referring exclusively to the affective realm were also used to describe physical sensations completely independently of affects. A "trembling of the heart" could be mentioned with explicit reference to affects, but it could also be one of several symptoms of a fever, for example. And in contemporary language use, it was not only a person who could get a "fright"; the heart too could at times be "frightened" in lay parlance. Even the blood could frighten. For example, a Breslau barber's apprentice related how he once in his youth "got a great fright in the nighttime" and made the mistake of

[91] P. Johannes Bozenhart, "Schicksale des Klosters Elchingen und seiner Umgebung in der Zeit des dreissigjährigen Krieges (1629–1645). Aus dem Tagebuche des P. Johannes Bozenhart". Ed. by P. L. Brunner, in: *Zeitschrift des Historischen Vereins für Schwaben und Neuburg* 1876, pp. 157–282, cit. pp. 238–239.

[92] Ulinka Rublack has argued along the same lines and has suggested moreover that the early modern modern body quite literally reacted differently from the way it does to today (Ulinka Rublack, "Fluxes: the early modern body and the emotions", in: *History workshop journal* 53 (2002), pp. 1–16).

drinking something soon after. He had no doubt that "due to this drink, the blood got frightened and became infected".[93] Michael Folckhamer, writing about the daughter of a neighbor who suffered from headaches, suggested a similar "frightening" of the menstrual blood: her suffering began in her youth when she had "a fright from a blaze"; when later she "got her *menstrua*, they did not occur at the proper times and only in very small amounts".[94] The examples of "Angst" ("fear") and "Bangigkeit" ("apprehension") exhibited most vividly the strong physical connotations of affective language in contemporary German. Certainly, one could experience "Angst" in the sense of being afraid of something back then. But *Angst* and *Bangigkeit* were physical sensations above all. They signified a feeling of pressure and "Enge" ("narrowness"), a term with an etymological connection to the German "Angst, to the English "anxiety" and to Latin *angustia* ("narrowness"). According to Juan Luis Vives, every "compression" of the heart in the case of illness, fear, anger, and desire was called "angustia", and he added, "This also happens often in the absence of an affect, when a thick humor burdens the heart".[95]

Early modern patient letters illustrate this impressively. Patients complained about "Angst" even without any perceptible connection to an affective experience, when during the course of a disease they suffered from shortness of breath or tightness in the chest. He had the greatest "Angst" in his chest without cease, complained Bernt von Luberstorff. It was so bad that he was "unable to draw a full breath", and sometimes "stitches at the heart" plagued him as well. These "complaints", he thought, probably came "from the stomach", from which morbid vapors ascended.[96] Similarly, a patient of Thurneisser gave an account of the severe coughing fits he suffered in the mornings. It fell from the head to the chest and lungs and made him feel so much "Angst" that he had to

[93] Staatsbibliothek Berlin, Ms. germ. fol. 424, fol. 309r-v, letter from Peter Müller, August 6, 1582.

[94] Ibid., foll. 365r-367v, letter from Dr. jur. Folckhamer, May 24, 1582.

[95] Vives, De anima [see note 10], p. 339.

[96] Staatsbibliothek Berlin, Ms. germ. fol. 424, foll. 82r-v and 85r, letter from von Luberstorff, March 13, 1582; on the notion of pathological vapors, which was very widespread in the popular medical lay culture of the time, see Stolberg, Homo patiens [see note 19], pp. 192–194 and pp. 222–239.

vomit with great pain and could hardly breathe.[97] Johan Kottwitz even asked Thurneisser for one of his costly urine tests—Thurneisser was well-known for diagnosing diseases by distilling the patient's urine—because he felt and "carried" such intense inhuman "Angst" and pain in the heart.[98]

In the same sense, people also spoke of an "interior fright" or "Bangigkeit" ("apprehension"). About his sick wife, the physician Philipp Fauchel said she had a strong toothache and became "often fearful and her hands tremble incessantly". She had told him that when she lay in bed, it came over her "like an interior fear" and dropped, it seemed to her, "suddenly in the stomach".[99] About a twelve-year-old boy it was said that he was "of a pale color, and had great pains in the body" and was often overcome by great "Bangigkeit", as if he were going to faint.[100] Ludolf von Closter related that he had had "strong complaints from head fluxes" which then fell "on the chest"; this gave him "great oppression on the chest" and "Bangigkeit" toward the heart and sometimes "not enough air". Thus, he asked Thurneisser to treat his head fluxes and to give him *confortativa* to "strengthen the heart".[101]

To underpin their claims that the chest, the heart, was the actual seat of the affects or passions, physicians and philosophers were thus able to refer to their readers' bodily experience as clear evidence. Thomas Wright asked: "for who loveth extreamely, and feeleth not that passion to dissolve his heart? who rejoyceth, and prooveth not his heart dilated? who is moiled with heaviness or plunged with paine, and perceiveth not his heart to be coarcted? whome inflameth ire, and hath not heart burning? By these experiences we prouve in our hearts, the working of passions

[97] Staatsbibliothek Berlin, Ms. germ. fol. 420a, fol. 87r-v, letter from Georg Traupitz, October 7, 1571.

[98] Staatsbibliothek Berlin, Ms. germ. fol. 420b, foll. 462r-464v, letter from Johan Kottwitz, March 19, 1575.

[99] Staatsbibliothek Berlin, Ms. germ. fol. 421b, foll. 106r-107r, letter from Fauchel, Elbogen, June 25, 1577.

[100] Staatsbibliothek Berlin, Ms. germ. fol. 420b, fol. 105r-v, letter from Balthasar and Moritz von Nischwitz, Feburary 3 [1574?], about an unnamed 12-year old boy.

[101] Staatsbibliothek Berlin, Ms. germ. fol. 420b, foll. 467r-468r, letter from von Closters, Wednesday after Misericordia, 1575.

and by the noyse of their tumult, wee understand the woorke of their presence".[102]

This shows at the same time that our contemporary, largely metaphorical understanding of the heart as the seat of emotions is by no means based in a timeless anthropological constant.[103] Rather, it proves to be the result of a long historical development. It reflects assumptions—unchallenged for centuries—about the nature and significance of the heart as an anatomical part of the human body and its central role in affective phenomena.

Learned Theory and Lay Culture

When we compare the early modern learned theory of affects, especially as it was expounded by physicians, with the statements made by (educated) laypeople, we find many commonalities. Physicians and laypeople both considered the affects to play a central role in matters of health and they both attributed a broad range of different disease episodes to strong, especially negative affects. As in many other areas of the medical, somatic culture of the early modern period, physicians and laypeople shared the basic notions. However, striking differences can be noted in some respects. They merit further investigation, including with regard to the specific early modern understanding of how affects were to be controlled.

First, physicians and laypeople alike believed that the affects as such not only had an effect on the body but were themselves largely of a physical nature. While vastly different positions were held, the medical literature by and large agreed that the soul played an important independent role in affective phenomena, even if the proper seat of the affects was the heart. Laypeople by contrast almost exclusively described the affects in somatic, bodily terms. They referred to an independent, halfway clearly defined psychological realm of "matters of the soul", at best in vague terms only. This was still the case in the eighteenth century. The only

[102] Thomas Wright, *The passions of the minde*, London 1601 (repr. 1973), p. 61.

[103] As much as I welcome their desire to give more weight to the physical, bodily aspects in philosophy, I am very critical of the claims of Hermann Schmitz and other representatives of the socalled "phenomenological" tradition that posit a universal experience of the body, rather than one that is—like that of these authors—specific to the respective time and culture (see e.g. Hermann Schmitz, "Leibliche Quellen der Herzmetaphorik", in: *Das Herz im Kulturvergleich*, ed. by Georg Bekemer and Guido Rappe, Berlin 1996, pp. 12–21).

change was that their descriptions increasingly used the language of the new medical model of a pathological and pathogenic irritability and sensitivity of the nerves. People now experienced affects and their influence mostly as a very literal "irritation" or "destabilization" of the nervous system, to which they were just as "passively" exposed, however, as to the movements of the blood and vital spirits.[104]

Second, physicians emphasized without cease the necessity of moderating, of controlling one's passions. This was a central message in their dietetic writings.[105] Thus, medicine once again acted as a moral authority that helped assert accepted social norms and Christian values and advance with medical arguments those processes that Norbert Elias has described as the "process of civilization".[106] If ethical and religious considerations were not reason enough to restrain oneself and control one's passions, then one could at least do it for the sake of health. However, lay people's ego-documents about daily life show that there was very limited acceptance of the physicians' medical advice. Dietetic rules of conduct were largely applied in times of sickness, as an important component of therapy. Furthermore, especially in the case of the affects, the physicians' recommendations were quite out of touch with everyday life. What was one supposed to do with an advice like, "those who wish to protect themselves from such evil, let go of your fear".[107] Ego-documents indicate a much weaker belief in the possibility of controlling one's affects and therefore in the possibility of actively guarding against the morbid influence of the affects. Certainly, it was sometimes possible to shield fellow human beings from pathogenic affects. Efforts were made to protect pregnant women from frightening experiences, for example, by sparing them the sight of an epileptic having a seizure. People also commonly withheld a negative prognosis from seriously ill family members for as long as possible, so as not to aggravate the course of the disease with fear, fright, or sorrow. But ultimately, especially where their own affects were concerned, laypeople described themselves as being largely at the mercy

[104] For a more detailed discussion see Stolberg, Homo patiens [see note 19], pp. 215–260.

[105] In this sense, Beverwijck, Schat, [see note 35], p. 26, for example, declared "de gerustheyt des gemoets het beste middel is, om langh in gesontheyt te leven.".

[106] Norbert Elias, *Über den Prozeß der Zivilisation. Soziogenetische und psychogenetische Untersuchungen*, 2 vols, 6th edn, Frankfurt 1979.

[107] Fabricius, Consilium [see note 37], fol. 459v, quoting Paré.

of their affects—presumably one reason being that they, more so than the physicians, understood affects primarily as bodily phenomena. It seems that in the lay experience much more than among physicians, the "affects" (from the Latin "afficere", "to have an impact on") or "passions" (from "pati", "to suffer") were perceived quite literally as something that was "suffered" or "put on a person".

Third, much more clearly than the physicians[108] some laypeople pointed at the disastrous consequences of "controlled", suppressed affects. For example, Willibald Pirckheimer, suffering from podagra, complained that while he had put up with injustices and animosity with a magnanimity of heart, he had not been able to prevent "their incessant and bitter stings from boring deeply into his heart".[109] In 1714, a patient ascribed his flatulence, toothache, and powerful rising "heats" to the fact "that I have been living for 20 months with a secret anger and have never been able to vent my passions".[110] And in the eighteenth century, Mme de Moncharlé was convinced that the "violent restraint" she had imposed on herself at the time of her wedding thirty years previously was largely responsible for her present breathing difficulties, flatulence, and cramps.[111] As we can see, not only strong affects but also the moderation or control of the passions could, in the experience of

[108] There are two noteworthy exceptions in the English debates. Francis Bacon (who was not a physician, however) found that an "inward conceived ioy, sparingly vented, doth comfort the heart more than a vulgar immoderate expression of ioy". In the case of wrath, however, he came to the conclusion that it consumed the spirits, when it was not vented, while their warmth was strenghthened when it was not restrained (Francis Bacon, *The historie of life and death. With observations naturall and experimentall for the prolonging of life*, London 1638, pp. 171–173); in particularly vivid terms, Timothy Bright described the immediate, bodily effects of anger, though without discussing the pathological effects: the upwelling of heat which anger caused in the area of the heart, was particularly intense, he found, "if it be not deliuered by word and deede, whereby liberty is giuen for the passion to breake forth, which restrained in any sort, breedeth an agony of such feruency, as it may resemble the scalding of a boyling chaldron not vncouered, or an hote fornace closed vp in all vents" (Bright, Treatise [see note 65], p. 88).

[109] Willibald Pirckheimer, "Autobiographie", in: W. P. Eckert and Christoph von Imhoff, *Willibald Pirckheimer. Dürers Freund im Spiegel seines Lebens, seiner Werke und seiner Umwelt*, Cologne 1971, pp. 125–134, cit. p. 132; see also Emil Reicke, "Willibald Pirckheimer und sein Podagra", in: ibid., pp. 184–202.

[110] UB Erlangen, Ms. 1029/I, foll. 658–659, October 1, 1714, illegible signature.

[111] Bibliothèque Cantonale et Universitaire, Lausanne-Dorigny, Fonds Tissot, IS 3784, undated letter by Mme de Moncharlé.

laypeople, be a great burden on one's health. At first glance, this might appear as an anticipation of the modern theories of psychoanalysis. Again there is a striking somatic element, however. There are striking parallels between the belief in the harmful consequences of withheld affects and the fear—ubiquitous in all lay testimonies—of the consequences on health of the "obstructions", "blockages", and "suppressed excretions". In a sense, affects here appeared like harmful morbid matter that by all means had to be removed from the body. Here, too, we can discern a tension between the medical insistence on controlling the affects rationally and the subjective lay perception of the affects as primarily physical in nature, as something external that befell a person who was more or less passively at their mercy and of which one had to rid oneself, like morbid matter.

Lukas Cranach's Representations of *Melancholia*, and the Medicine of His Time

At least four different paintings by Lukas Cranach that represent melancholy are known today. They all show *Melancholia* as a female, winged figure and were clearly inspired by Albrecht Dürer's famous engraving. Cranach added two new major elements, however: little children who are peacefully playing, and dark clouds or trails of smoke with naked women riding on animals, presumably witches, and other demonic figures. Historians have convincingly interpreted these demonic elements in the light of Luther's notion of melancholy as a "devil's bath". As this paper argues, they have missed a major point, however. The demonic figures in the dark clouds gave visual expression to the contemporary medical understanding of melancholy as a disease, rather than a melancholic temperament. The disease "melancholy" was characterized by delusions, by fury and frenzy or else despondency and was not attributed to a predominance of black

---

First published in German as "Lukas Cranachs 'Melancholia'-Darstellungen und die zeitgenössische Medizin", in: *Medizin und Sozialwesen in Mitteldeutschland in der Reformationszeit*, ed. by Stefan Oehmig, Leipzig: Evangelische Verlagsanstalt 2007, pp. 249–271. Translated into American English by Logan Kennedy and Leonhard Unglaub.

M. Stolberg, *Medicine and the Body in Early Modern Europe*, Palgrave Studies in Medieval and Early Modern Medicine,
https://doi.org/10.1007/978-3-031-84556-7_3

bile in the body but to morbid, adust, blackish humors that were generated from one of the four natural humors. They released dark vapors and fumes which ascended toward the head, disturbing the workings of the brain and causing delusions. This, the paper argues, was the idea behind the dark clouds in Cranach's paintings. The melancholic fumes rise toward *Melancholia's* brain. They make her believe that there are demons around her and prompt her to join them, to carve a stick for the ride to the Venus mountain, while the children around her play unperturbed.

Over the course of the centuries, "melancholia" has inspired a fair number of pictorial representations. Albrecht Dürer's famous engraving *Melencolia I* (Fig. 1), created in 1514, comes to mind most prominently. It is one of the best-known works in the history of Western art.[1] The subject was explored more extensively still by the Wittenberg painter Lukas Cranach the Elder. Four representations of "melancholia", produced in his workshop between 1528 and 1533, survive, and he possibly painted one or two more.[2] Cranach's *melancholia* paintings have not been given the same degree attention by art historians as Dürer's *Melencolia I*; one might say the famous model has overshadowed Cranach's representations of the subject. Even in Klibansky, Panofsky, and Saxl's comprehensive analysis of the art and cultural history of melancholy—their *Saturn and Melancholy,* which remains authoritative today—merely a few pages are dedicated to Cranach's paintings, and in the revised and expanded German translation of 1990, Klibansky provides only a short appendix as supplementation.[3]

[1] Raymond Klibansky, Erwin Panofsky, and Fritz Saxl, *Saturn und Melancholie. Studien zur Geschichte der Naturphilosophie und Medizin, der Religion und der Kunst*, Frankfurt 1992; Peter–Klaus Schuster, *Melencolia I. Dürers Denkbild*, 2 vols, Berlin 1991.

[2] Today, these paintings are in the Statens Museum for Kunst in Copengagen, in the Musée d'Unterlinden in Colmar, in the National Gallery in Edinburgh (on loan), and in a private collection. Dieter Koepplin and Tilman Falk, *Lukas Cranach. Gemälde, Zeichnungen, Druckgraphik*, 2nd edn, vol. 1, Basel and Stuttgart 1974, pp. 96–97, moreover quote a report by Beatus Rhenanus who saw a painting by Cranach with children playing and prancing when he visited Raimund Fugger in Augsburg. This may have been the 1528-painting today in Edinburgh. A couple of decades ago, also a black-and-white photo of a "Melancholia"-painting, dated 1524, emerged. At that time, the painting itself seems to have been in the possession of an arts dealer but its current whereabouts are unknown and it cannot be ascertained that it was truly painted by Cranach (cf. Klibanski, Panofsky, and Saxl, Saturn [see note 1], p. 565 and table 153.

[3] Klibanski, Panofsky, and Saxl, Saturn [see note 1], pp. 533–536 and pp. 563–658 (appendix).

Fig. 1 Albrecht Dürer: Melencolia I (1514)

Art historians have generally agreed that Cranach adopted important pictorial elements from Dürer, first and foremost the winged female figure, and also that he chose a somewhat similar composition,[4] all while countering the heightened humanist exaltation of learned melancholy with a considerably darker, negative understanding of melancholy, even casting it in a demonic light. Apart from several ambiguous elements and unanswered questions regarding some details, this interpretation is essentially coherent and helpful. However, from the vantage point of my own discipline, the history of medicine, I consider it incomplete. In particular, it seems to me that not enough attention has been paid to the medical connotations of Cranach's representations of melancholy. In the following, I will endeavor to show that the medical cannot be reduced to the old doctrine about the effects of excessive black bile in the body. Rather, Cranach gave expression to the then dominant understanding of melancholy as an illness in its own right, characterized by specific symptoms, an illness that was caused sometimes only, but certainly not always, by black bile. Vapors and trails of smoke that rose up inside the body played a key role in this conception and this, I believe, explains the true meaning of the large, dark cloud which, compared to Dürer's print, is the most striking new element that characterizes all extant versions of Cranach's *Melancholia* paintings.

Melancholy in Cranach and Dürer

Despite some controversy regarding specific points, scholars today still largely agree with Saxl, Panofsky, and Klibansky that Dürer created a predominantly positive image of melancholy. In so doing, he took a decisive step away from the negative evaluation that had been traditional since the Middle Ages. According to this negative understanding, melancholic people, controlled as they were by black bile, tended to exhibit properties that strongly resembled those sometimes identified with the vice of

[4] Pietro Scarpa, "Carpaccio: identificazioni e proposte II. Le dame veneziane ovvero: la Melancholia, un precedente per Dürer e Cranach", in: *Arte-documento* (1999), pp. 143–149.

acedia in the medieval tradition.[5] They were anxious, gloomy, and sluggish. They avoided company and were often deceitful, distrustful, and miserly.

Dürer by contrast took up an alternative interpretive tradition that had its roots in the pseudo-Aristotelian *Problemata*. Here, the question was asked, "Why do all exceptional men in philosophy or poetry or the arts prove to be melancholics?" And the answer was that black bile shaped an individual's personality, and if it was well-tempered, it could give those who mastered it exceptional abilities. If, however, someone's black bile ran too hot, there was a risk of rage and ecstasy, and if it ran too cold, one became miserable and might consider suicide.[6] In the late fifteenth century, the Florentine physician and philosopher Marsilio Ficino had brought renewed attention to this view with his work *De triplici vita*.[7] He described the melancholic condition as a burden and a danger but also as a distinction, even as the characteristic constitution of the exceptional scholar, and he connected this doctrine with neoplatonic and astrological ideas: the melancholic scholar was under the special influence of the planet Saturn, and his qualities became more pronounced still, if he—like Ficino himself—was born under the influence of Saturn and was therefore a child of Saturn. This was to say that the planet could have a direct influence on life on earth, specifically on people's physical and mental constitutions through the spirits it emanated. In this way, and despite its generally malefic influence, it became a wellspring of outstanding intellectual abilities. According to the interpretation of Saxl, Panofsky, and Klibansky, Dürer gave expression to this view in many ways. Holding a book and a pair of compasses, the figure of melancholy is deeply engrossed in thought. She is surrounded by symbols of intellectual activity, such as a sphere and a polyhedron, as well as by tools, which possibly refer to the influence of Saturn on the respective trades. The radiant celestial body in the upper left corner must be understood as Saturn.[8]

[5] Stanley W. Jackson, *Melancholia and depression. From Hippocratic times to modern times*, New Haven and London 1986.

[6] Pseudo-Aristotle, *Problemata physica*. Transl. by Hellmut Flashar, Darmstadt 1962, pp. 250–261.

[7] Marsilio Ficino, *Three books on life*. Ed. by Carol V. Kaske and John R. Clark, Binghamton, NY, 1989.

[8] Rejecting this interpretation, Schuster, Melencolia I [see note 1], pp. 108–115, points at the medieval tradition of health regimens and representations of the temperaments. In

Comparing this with the representation of melancholy in Cranach's work, we find not only several similarities but also striking differences, while all four surviving versions from Cranach's workshop are quite similar to one another.[9] They all show a winged[10] female figure on the right, draped in a long dress and sitting in a room. Not appearing especially sad or gloomy, she whittles a thin stick and looks out in front of her, without clearly focusing on anything in particular. In the best-known version of 1532 (Fig. 2), today in Colmar, there is a sphere in front of her and a sleeping dog as well as a table with a fruit bowl and a lidded vessel, presumably a wine chalice.

The middle ground additionally shows two partridges and four small, putto-like children, with one of them sitting on a swing, whose ropes appear to be coming down from out of nowhere. Behind them, the view opens on a landscape with a castle on a castle rock and a dark cloud populated with strange, seemingly demonic animals and figures. Predating this work, Cranach's earliest extant *Melancholia* painting, completed in 1528 (Fig. 3), still has tools lying on the floor next to the sphere, similar to Dürer's image. But there are no partridges. Two drinking vessels are on the table. The four children in this painting are busy waking or teasing a dog. A third version, dated to 1532 and held today in Copenhagen (Fig. 4) uses a more pronounced landscape format. The children, this time equipped with sticks and a hoop, play with an oversized sphere. There is no table with a fruit bowl. The dog lies, with now two partridges in front of it, in the large opening of a wall which, here too, affords a view of a

these, the melancholic temperament is sometimes linked to intellectual activity and represented by figures who are shown thinking or, in some cases, with a book before them. It remains doubtful, whether the figure in Dürer's "Melencolia I", following Schuster's argument (ibid., p. 403), should be seen as a "representative of virtuous wisdom", equipped with "all the features of artistic ability, fame, and success", which also does not seem to suffer "from a contradiction between divine inspiration and rational science". With his insistence on commonly known dietetic notions rather than esoteric, astrological and neoplatonic elements, Schuster certainly opens a plausible line of interpretation, however.

[9] On the following, see the catalogues of Cranach's works in Max J. Friedländer and Jakob Rosenberg, *Die Gemälde von Lucas Cranach*, Basel, Bonn, and Stuttgart 1979, and in Koepplin and Falk, Lukas Cranach [see note 2].

[10] At first glance, the 1528-painting seems to lack the wings. A somewhat larger copy of the painting, made in the late sixteenth century (Columbus Museum of Art, Ohio), in which the figure has wings, hardly leaves any doubt, however, that Cranach did paint the figure with wings, which went missing later only, when the painting was heavily cut on the sides.

Fig. 2 Lukas Cranach: Melancholia (1532, Musée d'Unterlinden, Colmar)

sparsely populated landscape. A closer look reveals a troop of armed men traveling through this landscape in the foreground. Once again, the upper left corner is filled with a dark cloud that is cropped by the edge of the painting and populated with demonic figures. The latest version, of 1533 (Fig. 5), is somewhat distinct from the others. The painting is populated by fifteen small children. Some of them are dancing, others are just waking up or, more likely, are lying down to sleep in front of two children who are making music with a flute and a drum.[11] The wall opening here is much smaller and has been moved to the upper right. Also, the somber cloud with the demonic figures now fills almost the entire breadth of the picture; it ends in a bearded man's head whose mouth seems to utter the word "melancholia" in the direction of the female figure.

Cranach's figure of melancholy, as opposed to Dürer's, does not indicate much in the way of mental efforts or abilities. While the sphere and the compass may be read as references to geometry, they can just as easily be seen as mere quotations, as a deliberate allusion to Dürer's famous engraving. Also, the compass only appears in the painting of 1528, and in the painting of 1532, the sphere is shown as an—admittedly oversized—toy. What we find in all of Cranach's extant *Melancholia* paintings, by contrast, are ghostly looking human and animal figures in a dark cloud.[12] These scenes are dominated by naked women, apparently witches, some of whom are riding animals backwards. They seem to indicate some wild, supernatural, demonic activity. In this context, the stick whittled by the figure of melancholy attains a more specific meaning: that of a magic stick, of the kind used by witches to ride through the air. Traditional knowledge had it that such a stick had to be debarked to avoid spirits nesting between wood and bark.[13] The Copenhagen *Melancholia* from 1532 (Fig. 4) shows in detail how the figure peels the stick, while in the other versions, she merely sharpens it. In all its ostensible ordinariness, the

[11] Cranach may have made a reference here to a lost melancholia-painting by Mantegna which showed 16 children or putti (Klibansky, Panofsky, and Saxl, Saturn [see note 1], p. 536.

[12] In Dürer's painting, such demonic elements can at best be identified in the strange creature in the title banner which combines features of a bat and of a dragon. The cloud is missing in the 1524-painting [see note 2] which is only known from a photograph and cannot safely be attributed to Cranach.

[13] Klibansky, Panofsky, and Saxl, Saturn [see note 1], p. 533.

Fig. 3 Lukas Cranach: Melancholia (1528, private owner, on loan in the National Galleries, Edinburgh)

Fig. 4 Lukas Cranach: Melancholia (1532, Statens Museum for Kunst, Copenhagen)

Fig. 5 Lukas Cranach: Melancholia (around 1533, private owner)

act of carving would thus be a sinister clue that the melancholic female figure intends to take part in the demonic action.

Art historians have rightly emphasized these demonic elements. Interpreting the bearded man's head in the 1533 painting as Saturn, Günter Bandmann concludes that Cranach intended to paint a "picture showing the fate of children who are born under the sign of melancholic Saturn". The artist, according to Bandmann, gave expression to the "old idea that the melancholic had a connection with sorcery and the work of the devil.[14] Bandmann interprets the winged female figure as the "genius of melancholy", as a "goddess of fate endowed with magic powers", while the children dancing and making music unselfconsciously symbolize innocent mankind. They would "very soon be touched by the magic rod of the genius and fall to the demons who are approaching in the background as the entourage of the planet god Saturn".[15]

Dieter Koepplin points more directly to Luther's repeated assertion that melancholy was a "balneum diaboli", the devil's bath, as it was known to John Chrysostom and the medieval tradition.[16] Following Luther, Cranach had wanted to show melancholy not as the privileged burden of superior scholars but rather as a particular danger that was best countered with the spiritual joy of the pious Christian. Cranach located "the special nature of the melancholic one-sidedly in the area of diabolical magic and sorcery",[17] and accordingly, his *Melancholia* painting emits the fumes of "diabolical temptation".[18] Koepplin adds that his paintings were intended to oppose "mockingly" the "humanist pathos in Dürer's

[14] Günter Bandmann, *Melancholie und Musik. Ikonographische Studien*, Cologne and Opladen 1960, pp. 63–98, cit. p. 91 and p. 87.

[15] Ibid., p. 79.

[16] Koepplin and Falk, Lukas Cranach [see note 2], pp. 292–293; on the traditional notion of melancholy as a devil's bath, see also Werner F. Kümmel, "Melancholie und die Macht der Musik. Die Krankheit König Sauls in der historischen Diskussion", in: *Medizinhistorisches Journal* 4 (1969), pp. 189–209, cit. pp. 202–203. Kümmel also looks at the contemporary debate on the question, whether melancholy only made it easier for the devil to perform his sinister works or whether it was, as such, already the work of the devil.

[17] Dieter Koepplin, *Cranachs Ehebildnis des Johannes Cuspinian von 1502. Seine christlich-humanistische Bedeutung*. Diss. phil., Basel 1973.

[18] Koepplin and Falk, Lukas Cranach [see note 2], pp. 292–293.

famous engraving.[19] They were meant to say: "You oh-so-clever humanists who reason by means of astrological arguments and are fascinated by gloom are simply being tickled by the devil at times". This, according to Koepplin, was a message directed in part at Melanchthon, whose belief in astrological, planetary influences had grown stronger, especially in the years around 1530, and led him to oppose Luther to a degree.

Raymond Klibansky, too, sees something "witch-like and seductive" in Cranach's figure of melancholy. He claims that where Dürer considered Saturn's influence, Cranach, following Luther, saw the work of Satan. More recently, Charles Zika has offered a closer iconographic reading of the demonic figures in Cranach's cloudscapes. He interprets them as a confluence of several related traditions: there is, first of all, the topos of the "furious army", an often-violent nocturnal congregation of the souls who died an early death or were otherwise temporarily precluded from salvation: the souls of the hanged, the drowned, of people who committed suicide and so forth. Then there is the idea of witches riding through the night. And a third tradition, which likely comes to bear most, is that of the visit to the *Venusberg*, according to which men are lured into a seductive trap by witches. Here, we find allusions to sexual upheaval, a reversal of the gender order, and, when Catholic clerics are depicted, an anti-Catholic polemic.[20]

These interpretations neatly fall in line with an assessment of Cranach's oeuvre that is widely accepted today. Though Cranach did sometimes work for Catholic rulers,[21] he is by and large considered a leading proponent of a Reformed, Lutheran pictorial program. It seems to me, however, that these interpretations do not pay sufficient attention to Cranach's specific, medically shaped understanding of melancholy as an independent disease with a clearly defined pathophysiology, an understanding that

[19] Ibid., pp. 268–269.

[20] Charles Zika, "Cranach's melancholia paintings. Witchcraft and sexual disorder in the sixteenth century", in: *Geschlecht, Magie und Hexenverfolgung*, ed. by Ingrid Ahrendt-Schulte, Dieter R. Bauer, Sönke Lorenz, and Jürgen Michael Schmidt, Bielefeld 2002, pp. 227–272; idem, *Exorcising our demons. Magic, witchcraft and visual culture in early modern Europe*, Leiden and Boston 2003, pp. 332–374. The representation of Catholic priests who succumbed to seduction in the 1533-painting was already pointed out by Konrad Hoffmann, "Cranachs Zeichnungen 'Frauen überfallen Geistliche'", in: *Zeitschrift des Vereins für Kunstwissenschaft* 26 (1972), pp. 3–14.

[21] Andreas Tacke, *Der katholische Cranach. Zu zwei Großaufträgen von Lucas Cranach d. Ä., Simon Franck und der Cranach-Werkstatt (1520–1540)*, Mainz 1992.

gives meaning not only to the pictorial elements that are new in Cranach's paintings compared to Dürer's engraving, but also shows the overarching pictorial message of Cranach's *Melancholia* paintings in a new light.

Melancholia as an Illness

Beginning in antiquity, the term "melancholia" in Western medicine not only referred to black bile, one of the four natural humors, which, if dominant, gave rise to the temperament of the same name. The term "melancholia"—and this began with Rufus and Galen—also served as the central generic term for illnesses which today would be referred to as "psychiatric" illnesses. In Cranach's day, such afflictions also went by names such as *insania*, i.e. madness.[22] Melancholy, in this sense, was not the result of a mere imbalance of the four natural humors but was a specific disease. It was marked by more or less specific symptoms, such as anthropophobia, anxiety, and sadness, delusions, and sometimes rage. The most important special form of melancholy was called "lycanthropy". Lycanthropes had a predilection for graveyards and dead body parts, and they sought solitude and often suffered from ulcers of the legs.[23] The illness "melancholia" was serious enough to justify therapeutic interventions that could be quite drastic. A woodcut from the thirteenth century (Fig. 6), for example, recommends burning a hole in the scalp of the *melancholicus* using a cautering iron, and then trepanning the skull, that is, opening the skullcap.

Following this understanding, most sixteenth-century authors—though their accounts differed in many details—did not attribute the pathological changes in the body of the melancholic to a mere predominance of natural black bile. As the ambiguous definition of the term *melancholia*, signifying both a disease and a humor, suggests, there was some overlap. An extremely melancholic temperament, a strong prevalence of cold and dry qualities in the brain or a predominance of black bile, could lead to pathological melancholy, to madness. However, ever

[22] Useful survey in Jackson, Melancholia [see note 5]; see also H. Günter Schmitz, "Das Melancholieproblem in Wissenschaft und Kunst der frühen Neuzeit", in: *Sudhoffs Archiv* 60 (1976), pp. 135–162.

[23] Michael Stolberg, "Lykanthropie", in: *Der Neue* Pauly, vol. 15/1: *Wissenschafts- und Rezeptionsgeschichte La–Ot*, ed. by Manfred Landfester, Stuttgart and Weimar 2001, pp. 243–246.

Fig. 6 Anonymous: Trepanation to treat melancholy (thirteenth century)

since Rufus and Galen—and this is something that keeps being overlooked, even in medico-historical research—melancholy as an illness, as opposed to a temperament, was usually not ascribed to the mere predominance of natural black bile or the related primary qualities "cold" and "dry". Rather, a pathological alteration of humors which in turn set free vapors were thought to be the main cause. Black bile was not necessarily involved at all. Cranach's contemporary Jason Pratensis (1486–1559), for example, generally attributed an "affectus melancholicus" to spoiled, black, putrid blood, which sometimes showed admixtures of bile.[24]

Galen had already attributed the morbid, melancholic humor to a number of different causes. Fevers, but also by mourning and worries could promote its generation. But the principal cause were dietary errors: eating raw, indigestive foods, for example, was already pointed out as a

[24] Jason Pratensis, *De tuenda sanitate libri quatuor*, Antwerp 1538, p. 263.

cause by Galen's source Diocles. Strong dark wine, too, was dangerous.[25] This might help explain two new elements in Cranach's Melancholy that do not occur in Dürer's image, and which art historians have not quite been able to explain: the wine chalice or other drinking vessels on the table as well as the fruit bowl. Fruit was generally considered a food that was raw, cold, and relatively difficult to digest. Scholars have pointed out that the fruit bowl and the chalice remain untouched, which they see as a further indication of the figure's lethargy and passivity,[26] but I find this interpretation unconvincing. To show lethargy and passivity, one would expect a plate with served but untouched food, instead of a fruit bowl. As to the wine, the *Melancholia* of 1528 (Fig. 3) with its half-empty drinking glass tells a different story anyway.

In Cranach's time, disease theory as taught at universities often connected the notion of a melancholic matter with more concrete explanations of its genesis in the body and its pathological nature. In particular, overly heated, burnt humors, which could literally turn to ashes, were now considered by many authors to be the crucial cause of melancholy. Statements to this effect are not only found in the writings of physicians but also appear, for example, with Melanchthon, in Cranach's immediate environment; he explored the subject of melancholy at some length.[27] The association of melancholy with blackness was no longer primarily based on the hue of natural black bile, but on images of burning and charring.

The notion of "burnt humors" is to be taken very literally here. According to an old belief, an intense inner vital heat, the *calor innatus*, was at work in the body. As a kind of flame of life, it maintained all vital processes and was also responsible for the assimilation of food. In the Galenic conception, food was not digested the way it is in today's understanding, but was literally cooked. Like a kitchen stove, a person's inner warmth heated up the food in the stomach, the liver, and ultimately in the entire body. In this way, it transformed it from a raw to a "cooked" state, which allowed it to be absorbed by the different body parts. However, when humors became too heated, they burned like food in an overheated

[25] Galen, "De locis affectis", in: idem, *Opera omnia*. Ed. by C. G. Kühn, Leipzig 1822 (repr. Hildesheim 1964), vol. 8, pp. 1–452, cit pp. 179–193.

[26] Koepplin and Falk, Lukas Cranach [see note 2], pp. 292–293.

[27] Philipp Melanchthon, *Commentarius de anima*, Wittenberg 1540, p. 150.

pan. They lost their natural qualities. They became acrid and preternatural and caused disease.

Which humor specifically became so heated that it burned was of secondary importance and mostly manifested in the symptoms the patient suffered from. Sometimes it was the natural black bile that burned and turned into *melancholia adusta.* As Melanchthon pointed out,[28] this would most likely happen when the pores of the skin were closed or plugged—pores were considered an important excretory pathway at the time.[29] The notion of burnt black bile was particularly associated with that of a caustic "acrimony", whose effects became obvious in the typical skin ulcers of lycanthropes. Due to the heat and dryness that characterized it even in its natural state, the presence of burnt yellow bile best explained the raging that could be observed in some *melancholici,*[30] the so-called *mania*, which some authors, however, considered a separate kind of madness next to melancholy. As Melanchthon and others explained, burnt blood could in turn lead to forms of melancholy that were characterized by a relatively good, cheerful mood. Democritos the laughing philosopher was a well-known example. Ever since Avicenna, the notion of burnt phlegm had completed the quadripartite scheme.[31] It offered an obvious explanation in particular for forms of melancholy that were characterized by excessive sluggishness and sleepiness. While Melanchthon did not explicitly speak of burnt phlegm, he did very similarly consider phlegm that became mixed with burnt black bile to be the cause of forms of melancholy that were distinguished by unusual sluggishness. He claimed that he had personally observed a mentally disturbed man, a *mente errantem,* who slept all the time and could be woken only with music played on a lute.[32] Perhaps this is the source of the motif of the sleeping and music-playing children in Cranach's *Melancholia* of 1533, a motif that is missing in the earlier versions, however.

[28] Ibid.

[29] Michael Stolberg, *Homo patiens. Krankheits- und Körpererfahrung in der Frühen Neuzeit*, Weimar 2003, pp. 174–178.

[30] See, for example, the student notes by Johannes Hessus on a lecture by the Wittenberg professor Augustin Schurff in the 1540s, Universitätsbibliothek Erlangen, Ms. 1016, foll. 52–57.

[31] Avicenna, *Canon medicinae*, Venice 1608, p. 498.

[32] Melanchthon, Commentarius [see note 27], p. 150.

Melancholic Trails of Smoke

In light of these ideas about a specific melancholic matter that resulted notably from an excessive heating or even burning of humors, Cranach's cloud populated with demonic figures acquires a very tangible meaning. It was more than a pictorial device employed by Cranach to express that melancholics in general were threatened by demons or diabolical machinations. Had he wanted to say nothing more, he could have simply placed the demonic figures, the furious army, the riding witches, or the visit to the *Venusberg* in the landscape in front of the house. Why a cloud, which, in the version of 1533, is even clearly located inside the house and yet accommodates full-grown animals (Fig. 5)? Comic book or cartoon aficionados may be reminded of thought bubbles used by artists to express with words or images what goes on in the heads of their figures. Anachronistic though it is, this analogy, I believe, hits the crucial point: the demonic figures in the dark cloud must not be understood to depict "actual" witches in the external world. What we see here is not "a throng of witches, appearing at the behest of Satan, surrounded by a dark cloud and charging into the clear skies", as Klibansky says, following Koepplin.[33] Rather, we are observing a purely mental image, a phantasmagoria visited upon the inner senses of the melancholic. The melancholic believes he or she is seeing something that in fact only exists internally, namely in a specific, anatomically defined place: in the cerebral ventricles and the surrounding brain substance.

This is in line with the medical theories of the time: building on the doctrine of "burnt" melancholic matter, medicine provided a concrete, material explanation for the development of pathological mood changes and delusions. the gloomy, anxious mood of those suffering from melancholy and their diverse delusions were not only caused by the pathological, burnt humors in a general sense but also more specifically and decisively by the blackish sooty vapors or trails of smoke that rose up from those humors. This was especially, but not exclusively, thought to be true for the so-called hypochondriacal form of melancholy, whose origin lay in the upper abdomen; beginning with Galen, physicians distinguished

[33] Klibansky, Panofsky, and Saxl, Saturn [see note 1], p. 534; Zika, Exorcising [see note 20], p. 374, does conclude, however, that Cranach, with this cloud, wanted to represent a "specifically diabolical form of delusion".

three basic types of melancholy, depending on whether the disease manifested only in the head, whether it originated in the upper abdomen, or affected the entire body.[34] Galen wrote that just as sooty fumes or smoke or generally thick vapors rose at times from the abdomen to the eyes, clouding a person's vision, a fumy, sooty kind of smoke rose from the black bile up into the head in patients with hypochondriacal melancholy, where it affected the thought processes.[35] Later authors confirmed and expanded Galen's explanation. Alexander of Tralles, for example, held that some forms of melancholy resulted from the vapors that were released when there was too much blood in the body.[36] Ishaq Ibn Imran, whose interpretation was passed on by Constantinus Africanus and had a long-lasting influence on the Western understanding of melancholy, even held that the tears of the sad melancholic resulted came from the rising vapors that liquified in the head and pushed outward through the eyes as tear drops.[37] Avicenna, too, whose *Canon medicinae* dominated Western medicine until far into the period of Humanism, explained that, especially with the hypochondriacal form of melancholy, vapors and trails of smoke clouded and blackened the clear pure animal spirits. This darkening inside the brain terrified sufferers in the same way that most people were afraid of darkness outside.[38] In the late sixteenth and early seventeenth centuries, many authors principally connected melancholy in all its variant forms with cloudy, sooty, dirty vapors above all else.[39]

The idea of vapors and trails of smoke that collected in the brains of those suffering from melancholy was common knowledge among physicians in Cranach's time. It surfaced, for example, in a transcript of lectures titled *De melancholia* delivered by the Wittenberg professor of medicine

[34] Galen, De locis affectis [see note 25], pp. 179–193; Jason Pratensis, De tuenda sanitate [see note 24], pp. 268–271.

[35] Galen, De locis affectis [see note 25], p. 189 and p. 191; similarly, Guillaume Rondelet, *Opera omnia medica*, Geneva 1619, pp. 208–209.

[36] According to Jackson, Melancholia [see note 5], p. 52.

[37] Ibid., p. 58.

[38] Avicenna, Canon [see note 5], p. 52. Along very similar lines, most authors, according to Avicenna, attributed nightmares ("incubi") to vapors, which ascended from coarse, bloody, phlegmatic, or atrabiliary matter.

[39] Michael Kutzer, *Anatomie des Wahnsinns. Geisteskrankheit im medizinischen Denken der frühen Neuzeit und die Anfänge der pathologischen Anatomie*, Hürtgenwald 1998, p. 115.

Augustin Schurff (1494–1548). Schurff, too, attributed melancholy in its different forms to burnt yellow or black bile or else to burnt blood or phlegm. Just as people became afraid when they were surrounded by dark, gloomy air, so the *melancholicus* experienced fear on account of the black bile or a melancholic vapor ("vapore quopiam melancholico") in his brain. Schurff, incidentally, is an obvious source for Cranach's medical ideas.[40] He was Luther's neighbor and family physician and, like Cranach, was a regular guest in Luther's house. We may further assume that he likewise regularly frequented Cranach's apothecary shop,[41] which was common for physicians at the time. His daughter later married Cranach's son.[42]

These ideas about cloudy vapors can still be found today in metaphorical expressions such as "mentally clouded" or "a clouding of consciousness", or in the German word "Trübsal", which means "grief" or "misery" but derives from the word "trüb" for cloudy. For this very reason, it is important to point out that the ascending trails of smoke and vapors must not be mistaken for mere metaphors. Laypeople as much as physicians were convinced that, especially during the "concoction" of food and sometimes due to inflammation or local accumulations of heated or spoiled humors, real vapors and trails of smoke rose upward in the body. Much more than today, the body was understood to be permeable, which meant that humors and vapors could move around in it largely unobstructed by any anatomical barriers. This imagery of rising vapors was so firmly rooted in early modern medical culture that some patients believed they could feel the hot vapors rising in their bodies.[43] Whenever vapors were unable to leave the body quickly via the skin, such hot rising vapors almost inevitably accumulated under the hard skullcap. Unless it was artificially opened, they could only inadequately escape (see Fig. 6).

[40] Universitätsbibliothek Erlangen, Ms. 1016, foll. 52–57.

[41] Cranach obtained a pharmacy privilege in 1520. On Cranach's economic activities in Wittenberg see Heinrich Kühne, *Lucas Cranach der Ältere in Wittenberg*, Wittenberg 1973; Monika and Dietrich Lücke, "Lucas Cranach in Wittenberg", in: *Lucas Cranach. Ein Maler-Unternehmer aus Franken. Katalog zur Landesausstellung Festung Rosenberg, Kronach 17. Mai - 21. August 1994*, ed. by Claus Grimm, Johannes Erichsen, and Evamaria Brockhoff, Augsburg 1994, pp. 59–65.

[42] Ortolphus Marolt, *Epithalamion scriptum viro integerrimo, Lucae Cranach, et pudicissimae sponsae magdalenae Schurfiae. Anno 1551*, Wittenberg 1551.

[43] Stolberg, Homo patiens [see note 29], pp. 192–194.

In the brain, which itself was considered to be permeable or porous, and especially in the cerebral ventricles vapors and trails of smoke could wreak havoc on the rational soul. Its faculties—imagination, memory, and reason—were thought to have their seat mainly in the ventricles, not in the brain substance itself. And these faculties were crucially dependent on the healthy, natural constitution of the animal spirits, which could move freely inside the ventricles, functioning as the immaterial soul's direct instruments and indispensable connection with the material body. To be able to fulfill their tasks as the rational soul's direct instruments, they had to be pure and subtle to the highest degree possible. Even the vital spirits, which were formed in the heart from the most delicate parts of the blood, were highly subtle and mobile. In the brain, as animal spirits, they became clearer still, almost like light. Inevitably, their natural subtlety and purity was lost as soon as blackish vapors and trails of smoke from burnt humors entered the ventricles and began to mix with the animal spirits. They became cloudy, dark, even black, and in turn they literally clouded and darkened the mood and reason. Now, everything appeared in a dim, obfuscated light. Fear, caused by darkness, and sadness as the two most important symptoms of melancholy, were the result. At the same time, the clouded animal spirits were greatly diminished in their ability to complete their important task as the immediate instruments of the soul. This affected the imagination especially. As early as the thirteenth century, Johannes Actuarius had ascribed hallucinations to an animal spirit that was polluted (*coinquinatus*) by a bad mixture (*temperatura*) or by the vapors of some spoiled humor.[44] Some of the physicians of the late sixteenth and early seventeenth centuries even thought that rising melancholic fumes and trails of smoke could accumulate in the skull in such a way that cloud-like formations resulted. Those affected believed that they had certain shapes or figures in front of them when, in reality, these existed only in their heads. They might be lying still, while the sooty vapors in their body deluded them into thinking that they had transformed into a wolf and were attacking innocent lambs, children, and women. This served as a medical explanation for ferocious forms of lycanthropy, the werewolf

[44] Ioannis Zacharias Actuarius, "De actionibus et affectibus spiritus animalis, huiusque nutritione", in: idem, *Methodi medendi libri sex*, Venice 1554, pp. 321–399.

delusion, and for the confessions of alleged witches who declared that they had transformed into werewolves.[45]

The knowledge of the significance of morbid melancholic vapors and fumes for the development of melancholic delusions is found not only in medical literature in the proper sense. As early as the thirteenth century, Guillaume d'Auvergne, for example, explained to his readers in great detail the development of delusions that could be observed with melancholics among others: they originated in vapors and smoke that rose up to the brain and settled there for longer or shorter periods. At such times, those affected could believe themselves to be a king or even Christ—a very dangerous disease, he emphasized.[46]

Pathophysiology and Spiritual Welfare

I believe that it is precisely to this tradition of interpreting melancholy and delusions that Cranach's *Melancholia* paintings with their clouds populated by demons make reference. Cranach painted what was happening in the inner world of a person, the goings-on inside the skull of the figure sitting seemingly unmoved. Her powers of imagination are affected in specific ways by the melancholic matter and the rising, dark and gloomy vapors and trails of smoke. Nothing of the sort happens in the outside world. The dog and the children show no signs of being affected. They clearly do not perceive any demons. But the mental images, the delusions, have such a persuasive power over the melancholic female figure that she begins to whittle herself a stick to partake in the imaginary demonic action. It seems likely that the partridges as well—whose iconographic interpretation has not been plausibly explained to date—signify that the cloud represents a delusion.[47] After all, the *perdix* (partridge),

[45] Cf. Stolberg, Lykanthropie [see note 23]

[46] Guillaume d'Auvergne, "De universo", in: idem, *Opera omnia*, Venice 1591, pp. 561–1012, cit. pp. 827–830. Guillaume's works (d. 1249) were already accessible in print in Cranach's days.

[47] Zika, Cranach's melancholia paintings [see note 20], p. 231, interprets them as a reference to hunting and as a link between the interior space and the landscapes outside the house.

wrote Conrad of Megenberg in his influential *Book of Nature,* is "quite mischievous and untrue", and "deceives man with lies".[48]

Cranach did not depict just any delusions to show the effects of melancholy, however. Rather, in all four paintings, the melancholic delusions are characterized by demonic figures of one kind or another—even though he surely would have been familiar with other melancholic delusional ideas that had been handed down in history as topoi: the frequently mentioned belief that one was made of fragile clay, or the fear that the sky was falling. One can only agree with Zika, Koepplin, and Bandmann that Cranach made specific reference to the time-honored notion, also taken up by Luther, that melancholy was "the devil's bath".

Bearing in mind the contemporary understanding of the nature of the illness and its physical genesis, as sketched out above, the notion of *melancholia* as a bath prepared for the devil attains a much more concrete, almost materialistic meaning, however. As Stuart Clark has argued very convincingly,[49] there was a widespread consensus in early modern demonology that, when doing his devil's work, the devil had to be content with manipulating the things that existed in nature as cleverly as he could within the laws established by God in His creation. Without God's permission, he was unable to override the limits of these laws. For example, as Augustine had pointed out,[50] he was unable to actually transform humans into wolves or asses. But using extraordinary trickery, he was able to deceive their senses in such a way that they were unshakably convinced of having been transformed.[51]

The doctrine of animal spirits that were pathologically altered by cloudy melancholic vapors and trails of smoke offered an immediately

[48] Konrad von Megenberg, *Das Buch der Natur*. Ed. by Franz Pfeiffer (repr. Hildesheim 1962), Stuttgart 1861, pp. 215–216. It was also said of the partridge that it stole and hatched the eggs of other birds: see Arthur Henkel and Albrecht Schönes (eds), *Emblemata. Handbuch zur Sinnbildkunst des XVI. und XVII. Jahrhunderts*, Stuttgart and Weimar 1996, coll. 842–845. According to Konrad von Megenberg the partridge also had "a dry brain", however, which could indicate a connection with dryness as a characteristic quality of natural black bile in humoral-pathological terms. Moreover, it was believed to be particularly unchaste.

[49] Stuart Clark, *Thinking with demons. The idea of witchcraft in early modern Europe*, Oxford 1997.

[50] Augustin, *De civitate Dei*, book 18, ch. 18.

[51] Andrea Cesalpino, "Daemonum investigatio peripatetica", in: idem: *Quastionum peripateticarum lib. V*, 2nd edn, Venice 1593, foll. 145r–168v.

plausible explanation why melancholics were especially susceptible to such devilish delusions and why melancholy offered a superb opportunity for the devil to achieve the strangest things ("miriabilia moliri"), as Melanchthon's son-in-law Caspar Peucer put it.[52] All the devil had to do was strengthen and intensify the natural disease process in the body. He simply had to increase the heat of the natural humors or mass together the smoke that rose from them and perhaps add to the build-up of smoke by preventing the excretion of unclean humors and vapors that normally occurred with menstruation or via the skin. He furthermore had to ensure that these substances affected the imagination in specific ways and would produce delusions that corresponded to his intentions, perhaps even by condensing the vapors and trails of smoke in the cerebral ventricles into certain forms and figures. The *melancholici* affected in this way believed they were seeing and experiencing demonic action in the outside world or were partaking in it. For example, they were convinced, and some of them testified to it in court, that they had made a pact with the devil or had even had sexual intercourse with him, or else they said that they had visited the *Venusberg* or gone to the Witches' Sabbath with other witches. Yet everything had only taken place in their heads.

Interpreted this way, Cranach's representations of melancholy have much more to offer than a mocking commentary on the neoplatonic and humanist valorization of the melancholic scholar. Beyond this, they are pictorial manifestos of Reformed anthropology as a whole, and this fundamental importance helps us understand why Cranach explored the subject so thoroughly. What we must consider here is not so much the anthropology of Luther himself,[53] but rather, the way in which Cranach's paintings contain essential elements of the very much medically inspired anthropology and the doctrine of spirits that Melanchthon developed during those years and later explained in detail in his influential work *De anima*.[54] Its title invokes the honorable tradition of the

[52] Caspar Peucer, *Practica seu methodus curandi morbos internos, cum generalis, tum particularis*, Frankfurt am Main 1614, pp. 191–192.

[53] In his consolatory letters to the melancholic Carthusian monk Joachim von Anhalt, Martin Luther largely equated "melancholia" to "tristitia" (Ute Mennecke-Haustein, *Luthers Trostbriefe*, Gütersloh 1989, pp. 167–169).

[54] Melanchthon, Commentarus [see note 27]; idem, *Liber de anima*, Wittenberg 1552; cf. Jürgen Helm, "Die Galenrezeption in Philipp Melanchthons *De anima* (1540/1552)", in: *Medizinhistorisches Journal* 31 (1996), pp. 298–321.

scholarly commentary on Aristotle's theory of soul. Upon closer examination, however, Melanchthon's *De anima* reveals itself to be a textbook on anatomy and physiology based in the Galenic tradition. His message was unmistakable and explicit: medical knowledge about the structure of the human body and its internal processes was indispensable not only for physicians, who were concerned with physical wellbeing, but also with respect to any believer's spiritual welfare. After all, thinking and the affects—here including especially religious belief—always had a somatic basis. If that somatic basis became pathologically altered, this necessarily had effects on morality and strength of faith.

The quality of the vital spirits in the heart, which was the seat of the affects, and the animal spirits in the brain played a key role in this. Therefore, Melanchthon was convinced that not only every physician but every believer had to be knowledgeable about the spirits' activity and effects in the body and about their pathological changes or clouding. Due to their central significance for affective activity in the heart and for the imagination and for the faculty of judgment in the brain, the spirits were ultimately crucial to spiritual welfare. One had to understand how they worked if one wanted to succeed in mastering one's affects, in providing the right place for one's faith to dwell, and in protecting oneself against the devil's temptations. If the holy spirit mingled in the heart with the subtle, clear spirits, making them more radiant still with its divine light, as Melanchthon wrote later, then faith was bolstered and the longing for God burned stronger still.[55] This could succeed all the better, the clearer and purer the spirits were in the heart. The devils, on the other hand, could corrupt the spirits in the heart and brain with their breath; they could disturb the faculty of judgment and tempt people into committing the most heinous crimes.[56]

To ward off such demonic machinations, pious prayer and spiritual joy offered only one pillar of support—though an essential one. They had to be bolstered with clear knowledge of the dangers of the physical changes that threatened salvation, dangers that might be countered with an appropriate lifestyle and diet. Considering Cranach's paintings in this sense as

[55] Melanchthon, Liber de anima [see note 54], coll. 88–89.

[56] Ibid.: "Econtra ubi diaboli occupant corda, suo adflatu turbant spiritus in corde et in cerebro, impediunt iudicia, et manifestos furores efficiunt, et impellunt corda et alia membra ad crudelissimos motus, ut Medea interficit natos, Iudas sibi ipsi consciscit mortem".

pictorial representations of Melanchthon's Reformed anthropology, one might, though admittedly in a somewhat speculative way, attribute a pedagogical function to them. Cranach would have warned the viewer of the dangers that—for every person, not only the scholar—might spring from a lack of knowledge about the processes in the human body. If a person lived unsuspecting like the playing children,[57] and did not know about the dangers of ordinary dietetic influences such as drinking dark wine and eating raw fruit, then they risked that vapors and trails of smoke might develop in their body, leading them to believe in a false demonic reality and making them easy prey to the devil's malice. Out of ignorance and by following an unhealthy diet, people risked not only their health but stood to lose the most important thing they had: their salvation.

If one wanted to accept an alternative interpretation of Dürer's engraving, attempted by Peter-Klaus Schuster in his extensive study in 1991, this would even open up a direct parallel. Schuster reads Dürer's image not least as a warning about the dangers of being overly active intellectually at night. In an analogous way, Cranach's paintings with the fruit bowl and the wine vessels could be seen as warnings about the dangers of making dietetic mistakes, which, however, would threaten everyone suffering from *melancholia* and its typical delusions, not only the scholar.[58]

Whether Cranach actually intended to convey this kind of concrete message, and whether or not he even read Dürer's engraving as an analogous warning about the dangers of mental overexertion, must remain open questions. But there is no doubt in my mind that any meaningful interpretation of his Melancholies must consider the contemporary medical understanding of *melancholia* as an illness characterized by specific pathological humors, rising vapors, and trails of smoke, especially if one aspires to do justice to the cloud populated by demons, the most striking novelty of Cranach's depictions.

[57] The children in Cranach's melancholia paintings have been interpreted in widely different ways. Zika (Cranach's melancholia paintings [see note 20], p. 247), for example, regards them as a representation of "moral disorder". Bandmann (Melancholie [see note 14], pp. 63–93) views the children's dancing and making music, in particular, as symbolizing *vanitas* and as a warning against the consequences of a life that is governed by sensuality.

[58] Schuster, Melencolia I [see note 1], pp. 108–115; cf. note 8.

"My Aesculapian Oracle!" Patient Letters as a Source of a Cultural History of Illness Experience in the Eighteenth Century

Over the past years, the history of illness from the perspective of patients—how patients experienced and coped with illness—has garnered increasing attention. It is a history that sheds light on a central aspect of human endeavor to cope with the challenges of life. It also opens paths for an empirical assessment of macro-historical theories, for example about the effects of power and law, of discourse formation, and of disciplining institutions on actual human experience and action. Not least of all, it allows us to see more clearly the characteristics of our own modern medical culture and its historical contingency.

Extant letters, diaries, and autobiographies cover a wide range of subjects, including illness and medicine, though they do so with varying degrees of detail and clarity, depending on the significance the author attributed to health matters. Given this situation, the systematic analysis of such sources in the context of a history of patients has thus so far been

---

First published in German as "'Mein äskulapisches Orakel!'. Patientenbriefe als Quelle einer Kulturgeschichte der Krankheitserfahrung im 18. Jahrhundert", in: *Österreichische Zeitschrift für Geschichtswissenschaften* 7 (1996), pp. 385–404. Translated into American English by Logan Kennedy and Leonhard Unglaub.

M. Stolberg, *Medicine and the Body in Early Modern Europe*, Palgrave Studies in Medieval and Early Modern Medicine,
https://doi.org/10.1007/978-3-031-84556-7_4

limited largely to the very commendable, yet not fully representative analysis of printed sources.[1] The case histories written by physicians focus on health matters and reflect a wider range of patients from different social backgrounds; less known today, this literary form was widespread in the medicine of previous centuries.[2] Drawing on a collection of case histories written by an early eighteenth-century physician from the German town of Eisenach, Barbara Duden has described the contemporary understanding of the body and illness with exemplary vividness. However, her study at the same time demonstrates the difficulty of accurately distinguishing in these sources between the patients' own experience of their bodies and their illnesses and the perspective of the attending and reporting physician.[3]

In comparison to these genres, there is a largely neglected genre of sources that combines the focused descriptions of medical case histories with the immediacy of personal testimonies: patient letters.[4] From the early modern period up until the nineteenth century, it was a widespread

[1] Roy Porter (ed.), *Patients and practitioners. Lay perceptions of medicine in preindustrial society*, Cambridge 1985; Roy and Dorothy Porter, *In sickness and in health. The British experience 1650–1850*, New York 1988; iidem, *Patient's progress. Doctors and doctoring in eighteenth-century England*, Cambridge and Oxford 1989; Lucinda McCray Beier, *Sufferers and healers. The experience of illness in seventeenth-century England*, London and New York 1987; Claudine Herzlich and Janine Pierret, *Kranke gestern, Kranke heute. Die Gesellschaft und das Leiden*, Munich 1991; Jens Lachmund and Gunnar Stollberg, *Patientenwelten. Krankheit und Medizin vom späten 18. bis zum frühen 20. Jahrhundert im Spiegel von Autobiographien*, Opladen 1995; on the rise of the genre of (auto-)pathographies, which focus on the personal experience of illness, see Anne Hunsaker Hawkins, *Reconstructing illness. Studies in pathography*, West Lafayette 1993.

[2] Cf. Christa Habrich, "Pathographische und ätiologische Versuche medizinischer Laien", in: *Heilberufe und Kranke im 17. und 18. Jahrhundert: Die Quellen und Forschungssituation*, ed. by Wolfgang Eckart and Johanna Geyer-Kordesch, Münster 1982, pp. 99–123; Jutta Dornheim and Wolfgang Alber, "Ärztliche Fallberichte des 18. Jahrhunderts als volkskundliche Quelle", in: *Zeitschrift für Volkskunde* 78 (1982), pp. 28–43.

[3] Barbara Duden, *Geschichte unter der Haut. Ein Eisenacher Arzt und seine Patientinnen um 1730*, Stuttgart 1987.

[4] Habrich, Pathographische und ätiologische Versuche [see note 2] and Porter and Porter, Patient's progress [see note 1], pp. 76–78 briefly mention the value of the genre as a historical source; on the importance of patient letters for the history of homeopathy see Jörg Meyer, "... 'als wollte mein alter Zufall mich jetzt wieder unter kriegen'. Die Patientenbriefe an Samuel Hahnemann im Homöopathie-Archiv des Instituts für Geschichte der Medizin in Stuttgart", in: *Jahrbuch des Instituts für Geschichte der Medizin der Robert-Bosch-Stiftung* 3 (1986), pp. 63–79.

practice for patients (or else their relatives or friends) to contact a famous and often faraway physician and ask him to give his medical advice based on a thorough description of the illness in question. Exploratory research shows that such letters have survived in considerable numbers among physicians' personal papers and offer excellent insights into patients' perceptions and interpretations of illness. At no other occasion did men and women feel as driven to speak with such precision and detail about their bodies and their medical histories. After all, whether or not the faraway physician would be able to arrive at a correct diagnosis and offer suitable treatment depended on the exactitude and completeness of their descriptions. Moreover, the patients or those writing on their behalf often had a strong desire to communicate to the physician their own ideas regarding the kind of illness at hand and its correct treatment. Therefore, the writers' own interpretive grids are found in a quantity that case histories, autobiographies, diaries, and private letters rarely offer, and we can gain manifold insights about the complex interplay between the physicians' conceptions of disease and therapy and those of patients.

Of course, patient letters, much like the other ego-documents mentioned above, derive almost exclusively from members of the literate, which is to say educated classes; at least this is true for the more distant past. Granted, some eighteenth-century authors of letters had obvious difficulties with spelling and syntax, and others relied on the help of relatives or even a professional scribe. Yet, the lower end of the social spectrum from which the vast majority of patient letter writers was comprised largely by school teachers, simple rural clergymen, nuns, and mid-ranking members of the military.[5]

In the following, patient letters from the estate of the well-known Swiss physician Samuel Auguste Tissot (1728–1797) will help us examine more closely several central aspects of illness experience in their respective contexts in the second half of the eighteenth century.[6] Like hardly

[5] Important insights into the ideas about diseases and their treatment that were common among the rest of the population can be gained from the writings of contemporary observers, such as visitation reports by members of the clergy or the numerous medical topographies and ethnographies penned by physicians in the eighteenth and nineteenth century, as well as from the minutes of proceedings in inquisition trials or trials against unlicensed healers.

[6] Bibliothèque Cantonale et Universitaire, Lausanne-Dorigny, Fonds Tissot, IS 3784 (quoted, in the following, as Fonds Tissot); on Tissot's life and works see Antoinette Emch-Dériaz, *Tissot, physician of the Enlightenment*, New York 1992; Emch-Dériaz briefly

any other physician of his time, Tissot was well suited to attracting the letters of patients from all over Europe who sought his advice. He first drew attention to himself in a broader public context with his forceful advocacy for a wide application of the controversial inoculation against smallpox.[7] But in 1760, he laid the true foundation for his spectacular rise to fame as one of the best-known physicians in Europe with his book *Onanisme*, the expanded French version of his *Tentamen de morbisex manustupratione*, published in 1758.[8] His popular success came in 1761 with the publication of his *Avis au peuple sur sa santé*, an accessibly written medical guidebook that included a list of medications that were easy to apply.[9] Two further international bestsellers followed: *De la santé des gens de lettres et des personnes sédentaires* of 1768[10] and the *Essai sur les maladies des gens du monde* of 1770. These books were translated into many languages and reprinted numerous times until well into the nineteenth century—the latest and not yet comprehensive list names well over 500 editions.[11]

Tissot became one of the most sought-after physicians among the nobility and the educated middle classes. Emperor Joseph called him to the chair of medicine in Pavia; the Polish king tried to win him as his court physician. Patients of high standing from all over Europe traveled to Lausanne to be treated by Tissot. Contemporaries wrote that the great influx of patients made it difficult at times to find acceptable accommodation in Lausanne.[12] Tissot's fame also spread to more modest middle-class circles. Hundreds if not thousands of patients turned to him, most of

mentions these letters but their outstanding value for the history of patients so far seems to have been overlooked.

[7] Samuel Auguste Tissot, *L'inoculation justifiée*, Lausanne 1754.

[8] Idem, *L'onanisme ou dissertation physique sur les maladies produites par la masturbation*, Lausanne 1760.

[9] Idem, *Avis au peuple sur sa santé, Lausanne* 1761; later editions came out with the slightly varied title *Avis au peuple sur la santé*. H. C. Hirzel published a German translation under the title *Anleitung für das Landvolk in Absicht auf seine Gesundheit* (Zürich 1767).

[10] It was based on an inaugural speech which Tissot delivered in Latin in 1766; a rather free translation appeared already in the following years, as *Avis aux gens de lettres et aux personnes sédentaires*, Paris 1767.

[11] Emch-Dériaz, Tissot [see note 6], appendix II.

[12] Charles Eynard, *Essai sur la vie de Tissot*, Lausanne 1839.

them from France but also from German and Italian lands, from Holland and Belgium. A conspicuously large number of letters were written by women, especially those letters in which the advice-seekers were relatives of the patient. Only few of Tissot's replies have survived, but many of the patient letters bear brief notes in his hand, most often giving the presumed diagnosis and the dietetic and therapeutic measures he had recommended.[13]

Patient-Centered Medicine

Despite the myriad diagnostic and therapeutic advances, modern medicine has been in crisis for several decades. Great concern is expressed about a lack of attention given to patients' subjective need for loving care and their desire to find meaning in their illness, about physicians who concentrate on individual organs, narrowing their focus on instrumental and laboratory findings instead of taking the "whole" human being into account, about the rift between professional jargon and patients' experience of their bodies and their illnesses. With all of this in mind, the medicine of the eighteenth century has received quite a favorable assessment in recent research. In the prevailing account, the medicine of that time assigned a key role to the personality of the sick individual, and the definitive basis for the physician's diagnosis and therapy was the patient narrative, the verbal description of individual complaints and the course of the disease from the perspective of the patient.[14]

Patient letters tell us about the weight given to patient narratives in eighteenth-century medicine, and they do so with considerable nuance. Remote medical practice was predicated on the idea that it was possible to do without an examination in person and to arrive at a correct diagnosis and treatment by relying solely on patients' descriptions. Patients were

[13] Presumably, these notes served to help a secretary of Tissot's to write a letter of response; cf. Antoinette Emch-Dériaz, "The non-naturals made easy", in: *The popularization of medicine 1650–1850*, ed. by Roy Porter, London and New York 1992, pp. 134–159, cit. pp. 141–142.

[14] Nicholas D. Jewson, "Medical knowledge and the patronage system in 18th century England", in: *Sociology* 8 (1974), pp. 369–385; idem, "The disappearance of the sick man from medical cosmology, 1770–1870", in: *Sociology* 10 (1976), pp. 225–244; Mary E. Fissell, "The disappearance of the patient's narrative and the invention of hospital medicine", in: *British medicine in an age of reform*, ed. by Roger French and Andrew Wear, London and New York 1991, pp. 92–109.

well aware of the particular responsibility that fell to them, and so their accounts reflect their efforts to give the faraway physician as comprehensive a picture as possible, one that included even the smallest detail that might possibly be relevant in establishing a correct diagnosis and therapy. The number and quality of excretions are described at length in these letters—from feces and urine to sweat and saliva to blood from bloodletting. Unpleasant or unusual sensations are noted as well, such as a bitter taste in the mouth, burning sensations, pain, feelings of bloating or tension in the belly, along with mention of insomnia, changes in appetite and sexual activity. Minute changes in everyday bodily functions are reported in great detail; unusual perceptions and the effects of earlier therapies are traced, all while showing how these occurrences relate chronologically to one another and how they correlate with the course of the illness and external events. More than once, the patience of the physicians reading such descriptions would have been tried, as many patients appear to have had difficulty ending their letters. They kept adding further facts and observations. Postscripts, in these letters, tend to be the rule rather than the exception, and sometimes a second letter followed the first even on the same day, containing information that was deemed essential but which had been inadvertently omitted.

Besides the description of the medical condition at hand, the letters often provide ample room for a biographical retrospective. Patients do not consider their condition in isolation but see it as part of, as resulting from, an ultimately lifelong disease process. They give in-depth descriptions of their many scars, of the specific instances in which their body was weakened and impaired in the course of their lives, deeming all of this essential in explaining their present complaints. A thirty-one-year-old legal consultant from Brescia, for example, begins his account by mentioning several smaller fevers, colds, and recurring headaches in his adolescence. He claims that they did not cause "any noticeable changes" of temperament. A violent fever he suffered in 1770, when he was twenty, left him with a "weakness" and a "dark melancholy", however, despite his successful treatment with bloodletting. This he considered the onset of a reversal of his circumstances, his "peripezie". Melancholy and sadness kept affecting him throughout the subsequent years, at times accompanied by excessive nervous tension ("stravaganze") and flatulence. This was followed, in 1773, by a putrid fever and significant weight loss. In 1780, a violent case of dysentery left him with severe flatulence and an extreme "weakness of the stomach". Finally, he caught a cold in the autumnal chill and, to his horror, kept seeing small admixtures of blood in his sputum. At the moment, he writes, he suffers from difficulty breathing, flatulence, a sense of pressure in his chest, twitching muscles, and an accelerated pulse.[15]

Here, one thing follows another in a sequence of symptoms which according to today's understanding are very different, but which, in the story of the author and his body, form a unified gestalt and, seen in their totality, become the subject of medical diagnosis and the basis of therapy.

What we begin to see beyond this emphasis on the biographical dimension is that the disease-related changes to one's inner and emotional life were commonly experienced and understood as a constitutive part of the disease process. Above all, despondency and sadness are mentioned in patient letters, along with the tendency to withdraw socially, but also barely manageable fits of anger, inner disquiet, haunting murderous fantasies, or else bouts of unexplained cheerfulness. Conversely, patients were convinced that strong affects had an active effect on bodily disease processes and that they were just as much a cause of disease as harmful physical and material influences.

In the 1970s, the British sociologist Nicholas Jewson advanced several theses to explain why during this period it was possible for a medical practice to become established and to prevail that was so patient-centered and apparently so much more in touch with the patient's individual personality than is the case today.[16] He argued that, in the eighteenth century, the position academically trained physicians occupied in the medical field was still precarious.[17] Their livelihood and their professional development often depended on the benevolence of a relatively small group of members of the elite, who usually enjoyed a higher social status than the physicians and who, in contrast to the rest of the population, preferred and were able to afford the medical care they offered. Jewson describes two central consequences of this situation. For one thing, he claims that it resulted in patients occupying a position in the physician–patient relationship that was extraordinarily strong compared to the situation today. If the physician wanted to secure his reputation and prevent losing his few

15 Fonds Tissot, letter from Orazio Gelisenti [?], February 20, 1782.

16 See note 14.

17 Claudia Huerkamp, *Der Aufstieg der Ärzte im 19. Jahrhundert. Vom gelehrten Stand zum professionellen Experten. Das Beispiel Preußens*, Göttingen 1985.

profitable patients to a competitor, he had to adapt his behavior and the medicines he prescribed to the needs of the patients as much as possible. For another, there were hardly any independent research institutions at the time. Rather, most researchers also practiced medicine. Therefore, the educated public, due to their superior position, was not only decisive for the career of individual physicians but also for the success or failure of medical innovations, and thus for development of medicine as a whole, for the topics and theories it would explore, and for the direction its practice would take.

When educated patients supported the proponents of certain theories or positions while withholding their support from others, Jewson claims, they effectively used their position of power to dictate a highly individualized medical understanding of disease that interpreted symptoms not as mere signs of pathological processes within the body, but considered their respective personal combination as the disease itself, an understanding that was particularly suited to their needs. Only by the nineteenth century, when medical care became increasingly professionalized and research moved from private practice to large city hospitals, did the situation change fundamentally. In the hospitals, physicians dealt almost exclusively with uneducated patients who ranked far below them in society. These were the preconditions for the new, objectifying perspective of modern medicine which, as Michel Foucault shows in his analysis, mainly considered specific organ pathologies.[18]

The patient's strong position in the physician–patient relationship as described by Jewson is largely confirmed in the patient letters. Not only when they looked at the course their disease had taken so far, letter writers took it for granted that they were competent judges of the diagnostic statements and therapeutic actions adopted by their physicians. Some openly reproached physicians, saying that their treatment had only made the disease worse: his mother had died at only sixty, one patient remarks tersely, "because during her final illness she was bled too much".[19] Even in their letters to the eminent authority of the famous Tissot, they show a remarkable self-confidence that follows on the heels of their introductory praises of his incomparable abilities. They suggest specific diagnoses

[18] Cf. Michel Foucault, *Die Geburt der Klinik. Eine Archäologie des ärztlichen Blickes*, Munich 1973.

[19] Fonds Tissot, letter from the Chevalier de Rotalier, November 1, 1771.

and actions. Pointing to their special sensibilities or to unpleasant side effects, they ask for a change in the therapy or simply change it on their own authority. Veritable haggling could ensue, for example about dietetic restrictions, about the necessity and dosage of unpleasant medicines, or about the time of day when they were to be taken. If the hoped-for improvement did not manifest quickly enough, even a physician such as Tissot had to contend with harsh criticism or a discontinuation of treatment on the part of patients.

Furthermore, Tissot's success as a medical author provides a good example for the career dynamics sketched out by Jewson. Without holding a prestigious position, lacking influential advocates among his colleagues, and living in the provincial Vaud area—a desert for science—he became one of the most famous and successful physicians in Europe almost exclusively by virtue of the great reception his writings received among the educated lay public.

However, the patient letters also reveal significant limitations in Jewson's explanatory model. It does offer a plausible explanation why at that time a medical practice came to the fore that offered space to the subjective sentiments and needs of patients, and why this type of medicine in the nineteenth century in the new conditions of hospital medicine yielded to a new medical perspective, even though, in hindsight, this new approach long brought only insignificant therapeutic progress. The configuration of the physician–patient relationship cannot, however, explain why, among the numerous possible medical models that prioritize the symptoms and subjective sensibilities of patients, certain models of illness predominated in the eighteenth century, with a specific imagery which has already been hinted at.

The Subjective Perception of Body and Illness

The patient letters mainly convey two disease theories that, in the late eighteenth century, focused patients's attention on the changes within their bodies and decisively shaped the perception of symptoms and disease: for one thing, we see different variants of the traditional doctrine of humors and temperaments, which had its roots in antiquity. According to the doctrine, the individual human being was shaped in body and character by a specific mixture of the four humors—yellow and black bile, blood, and phlegm—and by the qualities warm, cold, moist, and dry, which were connected to them. Diseases resulted from an imbalance of these humors and qualities, from pathological changes to them, or from an insufficient assimilation or harmful alteration of waste matter from food.

In the patient letters, people's perception of their bodies as informed by this model finds expression first of all in their preoccupation with their own bodily excretions—to an extent that, today, seems almost compulsive. The body appears as continually threatened by an inability to eliminate sufficiently harmful humors or "acrimonies", that might "throw" themselves upon an organ and cause the symptoms of a disease. A simple case of constipation is considered a serious warning sign, and great concern is expressed when the monthly cleansing of menstruation does not occur or unexpectedly stops. Men, too, experience periodic bleeding, such as from hemorrhoids or from the nose, helping the body excrete superfluous or bad humors. When such bleeding is delayed, this is noted with apprehension and considered the cause of spasms of the facial muscles, for example.

A qualitative or quantitative change of habitual excretions (such as night sweat) or the opening of new excretory pathways (for example weeping rashes or hemorrhoids) on the other hand point to an already active disease process in the body, which the organism is fighting by increasing its efforts to excrete. Concerned, patients write about bilious admixtures in their stool, about the unnaturally thick, viscous quality of blood from bloodletting, about the foamy or turbid appearance of their urine, or its "red" or pale color. Sometimes, an "unnatural" color or consistency of excretions gives the letter writer reason enough to sound the alarm despite the fact that he or she is otherwise in good health. Conversely, they sometimes share with great satisfaction the fact that despite their advanced age, their blood had been plentiful and "clear" during bloodletting.

The logical conclusion was to support the body in its efforts to excrete, using medical interventions such as bloodletting, leeches, laxatives, and emetics, menstruation-promoting baths, drawing poultices and sudorific medication, and in some cases even using procedures that led to a steady discharge of puss from a deliberately inflicted festering wound. Those who believe that patients would have quickly realized that such procedures were useless are sorely mistaken. In countless cases, those undergoing treatment wrote about a favorable course of the disease following bloodletting or the administration of purgatives. If treatment showed no positive effect, this could be explained by as resulting from the physician's choice of the wrong remedy or his having carried out the bloodletting at an unfavorable time or on the wrong vessel. A widower considered Tissot to be responsible for the death of his wife because, the man claimed, Tissot had closed an artificial ulcer on her leg without ensuring that the harmful humors could exit the body in another way. He was convinced that a deadly tumor had consequently developed from a local secretion of the humors.[20]

Further, the perception of one's physical and psychological identity was shaped by the concepts of humoralism: "I am of a bilious temperament", some people would write, or, "my blood is always heated and condensed". It is certainly possible that these patients were adopting statements made previously by attending physicians. Yet, the ease with which they drew on these concepts, connecting them with specific pathological tendencies that, in their experience, resulted from them shows the extent to which they had already made these interpretations and categories their own. There are many letters in which patients tell the faraway physician, who knows their body only through their descriptions, about their physical idiosyncrasies, and they tend to do this when they want to change their diet or therapy according to their ideas. "I noticed that showering has dried my blood", is one comment for example. And another patient writes that, "my bile tends to become heated", and since Tissot's treatment was fit to increase the amount of bile in the body, he fears that this might lead to local secretions of bilious "acrimonies".

Like humoralism, the second central disease model that we encounter in the letters was rooted in antiquity, but it was less coherent and the concrete form it took in the letters was essentially a product of the eighteenth century: the nerves, which permeate the entire body and connect it with the brain, control what happens in the body, and help explain the mysterious interactions between body parts and organs—interactions that used to be described by the neoplatonic notion of "sympathy". The concept of the "nervous diseases" that was derived from this constituted an eminently adaptive matrix for the classification of various physical and psychological changes. Furthermore, the "fibrous" structure of the nerves provided a loose connection with older, mechanistic conceptions of pathological slackening or excessive tensing of "fibers" in the body. Thus,

[20] Letter, dated March 15, 1792, quoted in William de Charrière de Sévery, *La vie de société dans la pays de Vaud à la fin du dix-huitième siècle. Salomon et Catherine de Charrière de Sévery et leurs amis*, vol. 2, Lausanne and Paris 1912, pp. 175–177.

many organ disorders could be ascribed to an affection of the nerves of the stomach, for example, followed by a slackening of the stomach and the accumulation of harmful undigested matter. Asthma was understood as the consequence of irritated pulmonary nerves. Tissot's famous contemporary, Robert Whytt, held that all diseases were ultimately nervous diseases of a sort.[21]

The nerves were also at the center of one of the most important and consequential discoveries of contemporary physiology, namely of the phenomena of irritability and sensibility. As these could be found in the living organism exclusively, their discovery undermined the previously dominant mechanistic explanatory models and furthered a rebirth of vitalism that started mainly in Montpellier, Tissot's place of study. Now, the body's organs and specific "forces", which could be at risk of becoming weak, were granted more attention once again. Because sensibility was no longer seen as a general property of the nerves but as a quantifiable state of the nerves in a specific organism, this model, analogously to the traditional doctrine of temperaments, ultimately led to the conception that people had a specific "nervous strength" and "sensibility" and that their susceptibility to pathogenic environmental influences therefore varied greatly.[22]

As in the case of humoralism, the patient letters reveal how close these conceptions were to people's experience and how they shaped contemporary perception. This was true for one thing of perceptions of lacking mental or physical vigor or increased irritability which still resounds in today's parlance when we say that someone's "nerves can't take it" or that they have "weak nerves". But above and beyond this, the letters generally feature the nerves as the very object of the author's subjective perception. They contain statements such as: "From this you can see that the nerves of my chest are very delicate", or, "my nerves have become excessively lengthened", or, "my entire nervous system has become weakened", or, "I suffer cruelly from the nerves".

21 Robert Whytt, *Observations on the nature, causes, and cure of those disorders which have been commonly called nervous hypochondriac or hysteric*, Edinburgh 1765; Tissot claimed that more than half of all chronic illness belonged to the group of nervous diseases (Tissot, Anleitung [see note 9], pp. 539–544 and pp. 644–645).

22 Sergio Moravia, "From homme machine to homme sensible. Changing eighteenth-century models of man's image", in: *Journal of the history of ideas* 39 (1978), pp. 45-60; Elizabeth A. Williams, *The physical and the moral. Anthropology, physiology, and philosophical medicine in France, 1750–1850*, Cambridge 1994.

At the same time, the new key position of the nerves and of sensibility, far more than the old doctrine of humors and temperaments, provided the basis for a "holistic" way of viewing the sick human being, for an integration of the world of feelings and affects and how they interact with all other physical phenomena. The following case may serve to illustrate this:

A twenty-four-year-old widow complained about a significant weakening of her stomach and entire nervous system ("genie nerveux"), accompanied by frequent attacks of the "vapors", of anxiety and panicky fear, along with further complaints that all pointed to an excessive "irritability". The widow considered her condition to have its roots in several previous diseases and, moreover, in her consuming grief. It had all started five years ago, shortly after her wedding, with severe colics and exhausting bilious vomiting. The distress and hardships of a two-year-long, unhappy marriage had dealt a blow to her nerves in the time that followed, bringing a deep melancholy. After her husband died three years ago, she had suffered from cramps for days, and a fever that lasted two weeks had left her completely emaciated. In the time after that, she had had choking fits, crying fits, violent headaches, and finally a bilious putrid fever. After a temporary respite, the "vapeurs" had recently increased again, especially during her menstruation, set off by strong impressions and emotions. The [bowel?] movements were increasingly like cramps, the nerves contracted from the fibers of her face all the way down to the womb. As of late, she went quickly, and in spite of herself, from attacks of crying and sobbing to intemperate laughter. Her heart sometimes beat in her throat. She felt strong pain below the ribs, accompanied by an urge to vomit. Despite her "rather good chest", she brought up blood several times and, one time, this had been followed by an eight-day fever.[23]

Today's readers, especially those who have medical training, often find it difficult to resist the temptation of making a diagnosis based on historical descriptions of illness such as these: "reactive depression" and "tuberculosis of the lung" would be two obvious possibilities. Yet, this example and others like it show very clearly how problematic retrospective diagnoses are. The description joins together into a cohesive whole are range of symptoms that are entirely unrelated in today's understanding, a uniting of dangerous symptoms like the expectoration of blood and

[23] Fonds Tissot, letter from Mme de Ruïß, née Mathieu, Strasbourg, May 26, 1777.

minor mental aberrations. Even before a person advances a clear interpretation of symptoms, their perception and weighing of physical and psychological changes emerges from culturally shaped suppositions about the construction and function of the body as well as pre-existing disease categories, most of which would have been advanced by physicians first and foremost.[24] With respect to our own history, too, one finding that has been repeatedly substantiated by cultural anthropologists needs to be taken seriously: even the seemingly most elementary bodily sensations such as pain are highly susceptible to cultural influences, and this is true even more so of the classification of subjective perceptions into complex disease entities which, in the extreme case of culture bound syndromes, can be found only in certain cultures to begin with, as exemplified by the Asian "wind disease", by "couvade syndrome", by male pregnancy—and by the "nervous diseases", "hypochondrias", and "vapeurs" of the eighteenth century.[25]

Meaning-Making, the Need for Control, Identity

Why exactly were large parts of the educated elites in the eighteenth century, while retaining the old concepts of humoralism, so willing and able to identify with the paradigms of "sensibility" and the related "nervous diseases" and—certainly we can think of Jewson here—to help them emerge so powerfully? Our necessarily somewhat speculative answer is: the new concepts superbly expressed the contemporary educated elites' self-image and their need for control.

[24] Only a few, exceptional letters describe bodily changes and sensations in idiosyncratic terms, which do not immediately make sense within the framework of current medical theories at the time. Patients wrote, for example, that they felt something like running through their vessels or an "effervescence" or "fermentation" of their blood. Even here, we can trace allusions to concepts from older medical traditions, however, such as the Paracelsion notion of "tartaric" diseases and iatrochemical ideas of "fermentations" in the body.

[25] Irving Kenneth Zola, "Culture and symptoms", in: *American sociological review* 31 (1966), pp. 615-30. The possibility of distinguishing the "biological" and the "cultural" by opposing "disease" and "illness", as has been done in English and American writing, remains controversial. For a refreshing account of the sometimes astonishing differences regarding preferred explanatory models, diagnoses, and therapeutic practices in various national cultures today, see Lynn Payer, *Medicine and culture. Notions of health and sickness in Britain, the U.S., France and West Germany*, London 1989.

In light of the new ideal of "naturalness", stylizing oneself as an exceptionally sensitive noble person or member of bourgeois society gave vivid expression to a growing discomfort regarding the shadow side of modern and in particular urban living: the whirlwind of encounters and impressions one experienced in the increasingly hectic cities; the forced cohabitation with throngs of people in limited space; the artificiality of daily rhythms of life; the constraints of conventional social life; the economic competition; but also an idleness that was perceived as devoid of meaning. As early as 1733, Georg Cheyne stated that nervous diseases had been perpetually on the rise, and Tissot's writings can be read in large part as medical variations on the theses proposed by Jean-Jacques Rousseau, for example when he declared his time to be an age of weakness, and considered the illnesses of the wealthy and educated to be largely caused by an estrangement from the simple, healthy lifestyle of the farmer.[26] The letters show how Tissot's concepts of excessive "sensibility" as well as "nervous diseases" allowed his readers to lend medical meaning to even vague malaise or mood swings, feelings of emptiness and ennui, including hypochondria[27] and "vapors", which were certainly understood as physical ailments. In a time in which the aspiration for salvation in the hereafter was increasingly giving way to a utopia of perfect health, even relatively minor impairments assumed their own medical meaning and were seen as requiring attention and consideration.

As the concept of sensibility came to not only describe the quality of the nerves but also to signify a character trait, "sensibility" became a sign of elevated refinement and mental and emotional sophistication, and the diseases that came of it became a medium of self-assurance.[28] Already in their sensitivity to even minor influences on body and mind and by suffering from the resulting diseases, the dominant and upwardly mobile classes distanced themselves from the robust physical health of the lower classes, who were increasingly experienced as dangerous. These were

[26] This is true, in particular, for the *Avis au peuple* [see note 9] and the *Essai sur les maladies des gens du monde* (3rd edn, Zürich 1771).

[27] Samuel Auguste Tissot, *Abhandlung über die Nerven und deren Krankheiten*. Transl. by J. C. G. Ackermann, vol. 1, Leipzig 1781, p. 9.

[28] George S. Rousseau, "Cultural history in a new key. Towards a semiotics of the nerve", in: *Interpretation and cultural history*, ed. by Joan H. Pittock and Andrew Wear, Basingstoke and London 1991, pp. 25-81; similarly already, Esther Fischer-Homberger, *Hypochondrie. Melancholie bis Neurose. Krankheiten und Zustandsbilder*, Bern 1970.

people who were spared nervous diseases, but were thought incapable of higher spiritual emotions and were not by all appearances contributing to the progress of mankind. In this period, social differences were thus biologized as physical differences, just as the difference between the sexes was attributed—now more than previously—to a fundamentally biological difference.[29]

The outward expression of the extraordinary sensibility with which the elite classes thought themselves endowed took the form of the desire for an accordingly "gentle" medical treatment. Now it was important to forgo the usual drastic measures and generous bloodletting, pursuing instead simpler, "more natural" means that were especially befitting to the delicate organism of the patient.[30] In the prescriptions of Tissot, donkey's milk and water, for example, were raised to the rank of panaceas, and they were received as such by his patients. Cold and warm baths played a similarly central role, carried out privately at home or demonstratively in the public space of a classy health resort.

The conviction of one's own exceptional sensibility significantly enhanced the status of dietetics, the pursuit of a healthy way of living. The basic dietetic principle of moderation, as William Coleman has shown,[31] was tailored to suit the upwardly mobile bourgeois classes. This principle allowed them to distance themselves from both the "excesses" of the "decadent" nobility and from the "uncivilized" lower classes. And unlike the majority of their contemporaries, they generally had enough time and the financial means to put dietetic requirements into practice.

Not only this, but dietetics—as opposed to trusting in medical remedies, magical rituals, or pious prayer—put responsibility over one's health in the hands of the individual burgher. Thus, it accorded well with the bourgeois ideals of autonomy and self-determination and satisfied in a comprehensive manner a desire for control, both retrospectively in the

[29] Thomas Laqueur, *Making sex. Body and gender from the Greeks to Freud*, Cambridge, MA, and London 1990.

[30] The striking parallels with the later success, among the more educated sectors of society, of homeopathy in the nineteenth century—which was likewise perceived as "gentle"—remain to be analyzed in more detail.

[31] William Coleman, "Health and hygiene in the *Encyclopédie*. A medical doctrine for the bourgeoisie", in: *Journal of the history of medicine and allied sciences* 29 (1974), pp. 399-421.

form of knowledge about the mistakes one had made and proactively by way of following specific guidelines.[32]

Michel Foucault and the Paradox of Power

Unlike today, when the medical ideas laypeople have about the nature and causes of their illnesses correspond only minimally with the understanding of physicians,[33] late eighteenth-century patients felt that their own ideas about illness were largely in line with the dominant medical views. Already back then, a physician might have attributed an unspecific complaint in the upper abdomen to inflammation and an adhesion of the liver to the diaphragm—and this was, incidentally, well before the much-invoked birth of the new clinical view in early nineteenth-century Parisian hospital medicine.[34] But on the whole, a common language was spoken, and (educated) patients felt that their whole personality was appreciated and taken seriously. Diagnoses and treatments were comprehensively experienced as meaningful, and they had a place in a shared view of the world and humanity. There is no indication in the classes represented here of a forced medicalization imposed from above, a medical acculturation through a repression of hitherto preferred forms of the healing arts.

More than anyone else, however, Michel Foucault emphasized the omniscience and polymorphism of power. Power, as the central thesis of his later work in particular holds, does not only work through repression. Rather, in the "will to knowledge", in the creative production of meanings, of discourses, in the definition of something that people can recognize as "true", power penetrates much deeper into people's inner

[32] Tissot emphatically warned especially his affluent readers, who could afford calling a physician anytime, that they also had to do their part in maintaining and, if necessary, reestablishing their health (Tissot, Essai [see note 26]).

[33] On the characteristics of "medical lay theories" and their positiv impact on the process of coping with disease see Claus Bischoff and Helmuth Zenz (eds), *Patientenkonzepte von Körper und Krankheit*, Bern 1989; Hans-Eckhard Langer and Helmut Bormann, "Krankheits-Bild als Krankheit-(Be-)Deutung. Versuch einer Meta-Theorie der Laientheorien bei rheumatischen Erkrankungen", in: *Psychologie in der Rheumatologie*, ed. by Heinz-Dieter Basler and et alii, Berlin 1992, pp. 55-82; David Locker, *Symptoms and illness. The cognitive organization of disorder*, London and New York 1981; Arthur Kleinman, *The illness narratives. Suffering, healing, and the human condition*, New York 1988.

[34] Fonds Tissot, letter to the husband of sick Mme de Moutrand, February 19, 1754.

being than simple prohibitions do—into their most personal feelings and perceptions and behaviors; in fact, it penetrates into the constitution of the subject as such.[35]

With this in mind, the great attention given in the contemporary medical culture to the way (educated) patients subjectively experienced their bodies and illness takes on a new dimension. Without question, Tissot's "holistic" medicine as well as that of fellow physicians was particularly well suited to creating personal meaning in the face of a threatening disease, and to providing orientation by tying in a patient's personal life history. This presumed, however, that patients would be ready for comprehensive self-disclosure, ready to divulge even the subtlest perceptions and the most inner goings-on in their bodies and emotions. Insofar as such self-disclosure was made the subject of linguistic expression, it was at the same time objectified and to some extent subordinated to the shaping, normative power of the dominant discourses about the body which, as we have seen, already had a significant influence on how disease symptoms were perceived in the first place.

Further, the experience of meaning was inextricably linked with the hope for control. Not only in Tissot's writings did dietetics as a comprehensive medical art of living take on a crucial role; it was also a central subject in Enlightenment medicine as a whole. The message was unmistakable: health can be achieved through action. All that was required was a willingness to monitor and control all possible external influences and inner goings-on that could be harmful when moderation was not upheld. In particular, it was in the physicians' ability to keep a perilous lack of power and loss of control at bay for patients that this holistic, individualizing medicine also proved to be supremely suited to advancing the process of civilization described by Norbert Elias as an increased internalization of social norms and a control of natural impulses through self-constraint.[36]

[35] Expressed in programmatic terms in Michel Foucault, *Sexualität und Wahrheit*. Vol. 1: *Der Wille zum Wissen*, Frankfurt 1983; see also Foucault's oral explanations in Michel Foucault, *Dispositive der Macht. Über Sexualität, Wissen und Wahrheit*, Berlin 1978.

[36] Norbert Elias, *Über den Prozeß der Zivilisation. Soziogenetische und psychogenetische Untersuchungen*, 2 vols., 6th edn, Frankfurt 1979.

This becomes particularly evident in the example of that category of disease for which Tissot is most known still today, the *morbiex manus-tupratione*.[37] Much attention has been called to the repressive tenor of Tissot's anti-masturbation tract. The terror elicited by his description of the harrowing consequences of masturbation (and other sexual "misdemeanors" or "excesses") for the body and the soul is hard to top. And his depictions are followed up with an outright demand for parents and teachers to go so far as to carefully monitor even pre-pubescent children as a precaution. In a time in which the religious sanctioning of moral values was losing significance, medicine stepped in as a new authority that no longer foregrounded the "vice" of onanism itself, but instead its supposed physical and psychological consequences, so as to reinforce the existing sexual norms and ultimately, through a clear preference for heterosexual, conjugal intercourse, to strengthen the dominant social order.[38]

But when we only consider the repressive character of the contemporary discussion of onanism, we do not recognize the full extent of the creative, meaning-making, and action-guiding power of Tissot's explanations. In the context of a holistic conception of the body and illness, the inclusion of the sexual in medical guidelines for lifestyle makes complete sense—one may think of the strong emotions that accompany the act, the "convulsive" climax that affects the nerves, the slackening and relaxation that occurs afterward, as well as of the transcultural fears of losing a valuable (male) vital humor or, in Western tradition, the extremely delicate, ether-like *spiritus animales*. And in fact, patients accepted without any major reservations the medical significance of sexuality and thus the necessity of disclosure in this area. Certainly, words such as "shameful" and "confession" occur from time to time, and here and there relatives or friends take over the reporting, saying that the patient himself might possibly be withholding certain things. But far more patients write openly about their sexual experiences and "misdemeanors:" about traumatic first encounters, the fear of impotence, premature ejaculation, strict fathers, unloving or even sexually assaulting husbands, about torturous desire, or

[37] The title of the Latin text is more precise. It was not masturbation as such, as has often been assumed (see, for example, Tristam H. Engelhardt, "The disease of masturbation. Values and the concept of disease", in: *Bulletin of the History of Medicine* 48 (1974), pp. 234-248) but its effects on the body which were seen as pathological.

[38] See, for example, Ludmilla Jordanova, "The popularization of medicine. Tissot on onanism", in: *Textual practice* 1 (1987), pp. 68-79.

else a complete loss of desire, but also and above all, about a past or perhaps still ongoing penchant for self-gratification.

For some patients, Tissot's construction of the new disease class of the *morbi ex manustupratione* even provided a long searched-for explanation of present complaints. M. Belfontaine, for example, the 43-year-old son of wealthy parents, claimed that he had not been truly healthy for twenty years, that he suffered from constipation, winds, headaches, and "vapeurs". Added to this came a painful and very itchy skin disease. He said that his physicians, however, did not really want to believe him, and some of them had even treated him as an imaginary invalid and sent him away with a variety of diagnoses. He felt he was in the best hands with his current physician as this physician understood better than the others that his illness was a nervous disease and treated him accordingly. Yet, he was still not fully restored to health. Only his reading of Tissot's work on masturbation had made him think that it was possibly "this unfortunate pleasure, so widespread among young people" that was the true cause of his suffering. And he went on to explain that, though his physician said he certainly did not suffer from the spinal consumption described by Tissot, he himself had noticed that his memory and vision seemed somewhat consumed, that his feet were sweaty, and his urine flowed excessively and was sometimes "raw" ("cru"), sometimes red in color. Also, his sexual desire was low; his erections were weak and he ejaculated immediately whenever he "saw" a woman, which happened rarely enough. He thought that he had perhaps not indulged in masturbation more than others in his youth but perhaps it had nevertheless been too much of a burden considering the weak disposition of his fibers. Furthermore, he had started before puberty—something Tissot warned against emphatically—and, worried that he might exhaust himself, had stopped before ejaculating. Tissot, after all, had cautioned about the dangers of just stimulating oneself sexually. When he associated with girls, he had them gratify him with only their hand because he was afraid of his father, who would not have forgiven him had he contracted a venereal disease in his youth. Belfontaine remembered that Tissot had emphasized how much less dangerous the loss of semen was when it occurred during heterosexual intercourse. But in any case, he had already started the treatment suggested by Tissot for cases such as his. Begging pardon for his "embarrassing confession" ("honteuse confession"), necessary though it was to enable a full understanding of his ailment, Belfontaine asked Tissot for additional therapeutic advice.[39]

M. Belfontaine was not an isolated case. Other patients, too, focused first of all on their subjective degree of suffering, and Tissot's construct of a *morbusex manustupratione* gave them the framework in which to conceptualize it. Later, Tissot would even refuse to answer letters from patients such as Belfontaine, referring to his time constraints.[40] It may be surprising at first glance that patients, based on a theory of illness that is abstruse in our present understanding, were willing to seek the causes for their complaints in their own wrongdoing. Yet, modern empirical studies have shown that comparable phenomena continue to be observed: it appears that understanding oneself as partly responsible for one's illness is easier to bear for many people than the feeling of being powerless and at the mercy of blind fate.[41] How much more comforting it must have been to know that, more than anything, the illness reflected one's insufficient consideration for the particular sensibility of one's organism, for the delicacy of the nerves, for the sensibility of one's person, which one could henceforth protect better, giving it unceasing care.

As indicated here, the ability to fashion creative body images and medical explanatory models that ill people experience as meaningful based on their previous views and their changing social situation is generally a decisive factor in the establishment of medical discourses and their normative power. Conversely, this means: within the conditions of patronage, as summarized by Jewson, but also beyond this—that is, generally as long as patients are able to choose between different forms of medicine—the power of physicians goes only as far as society, or more specifically the ruling elites, are willing to allow. The professionalization of modern science-based medicine has been able to diminish but not to eliminate this freedom of choice. And so, today's modern science-based medicine, feeling pressure mostly from the more educated classes, has long been adopting alternative "holistic" practices, for example, naturopathy and homeopathy, integrating them in university curricula, and more and more physicians are offering these outsider approaches to treating patients

[39] Fonds Tissot, letter from Monsieur Belfontaine to Tissot, Paris, November 25, 1772.

[40] Samuel Auguste Tissot, *L'onanisme ou dissertation physique sur les maladies produites par la masturbation*, expanded 8th edn, Lausanne 1785, preface.

[41] See, for example, the contributions to Bischoff and Zenz, Patientenkonzepte [see note 31].

alongside established science-based treatments. By the same token, there are many indications that, since the middle of the nineteenth century and up until the most recent past, large segments of the educated classes have found meaning and orientation precisely in the science-based medical practice that today is criticized for de-individualizing and objectifying illness and the body. Presented with its increasingly technology-driven and specialized approach to the body, these groups have found more hope that their complaints may be controlled than in the many competing "outsider" approaches that have always tried to enlist supporters. The displacement in the nineteenth century of rural, "folk medical" healing culture, and its "irregular" proponents, accompanied as it was by at times massive repression by the state, is another story of course.[42] Acculturation and cultural change can occur in a variety of ways.

[42] See, for example, on the situation in Bavaria, Michael Stolberg, *Heilkunde zwischen Staat und Bevölkerung. Angebot und Annahme medizinischer Versorgung in Oberfranken im frühen 19. Jahrhundert*. Diss. med., Munich 1986, and the less negative assessment for Baden in Francisca Loetz, *Vom Kranken zum Patienten. "Medikalisierung" und medizinische Vergesellschaftung am Beispiel Badens 1750–1850*, Stuttgart 1993.

Cura Palliativa. The Idea and Practice of Palliative Treatment in Pre-modern Medicine (ca. 1500–1850)

Introduction

Over the past twenty-five years, medical practitioners and the public at large have come to increasingly appreciate the significance of appropriate palliative care for incurably ill and dying patients. Around the world, hundreds of hospices, palliative care units, and ambulatory palliative care services have been established. Driving the development are numerous expert associations and hospice societies, more than a dozen medical journals, and not least of all the World Health Organization and national health ministries. Still, there are major deficits in many places. Even in the wealthy industrialized countries, there continues to be a dearth of palliative care beds, and many physicians have insufficient knowledge in the field, especially with regard to pain management. Not only this, but medical staff show both open and hidden resistance vis-à-vis palliative care. It seems the idea behind it contravenes a widespread self-image in which modern medicine is defined by its ability to heal and preserve

---

First published in German as "'Cura palliativa'. Begriff und Diskussion der palliativen Krankheitsbehandlung in der vormodernen Medizin (ca. 1500–1850)", in: *Medizinhistorisches Journal* 42 (2007), pp. 7–29. Translated into American English by Logan Kennedy and Leonhard Unglaub.

M. Stolberg, *Medicine and the Body in Early Modern Europe*, Palgrave Studies in Medieval and Early Modern Medicine,
https://doi.org/10.1007/978-3-031-84556-7_5

lives. Measured against this therapeutic ideal, palliative care even today appears to many physicians and hospital administrations to be tantamount to admitting failure.

Amidst the current efforts and debates, there has been an increased interest in the historical roots of palliative medicine, not least of all among its supporters. In recent years, the developments that have taken place since the Second World War have been portrayed multiple times: the groundbreaking opening of a hospice for the dying by Cicely Saunders 1967 in London,[1] the opening of the first palliative care unit under Balfour Mount in Montreal[2] as well as the establishment of the first institutions of the kind in certain countries and the further spread and institutionalization of palliative care.[3] It is quite uncommon, however, for the authors of such descriptions to look further into the past. Searching for the origins of today's hospices for the dying, researchers have identified precursors in the old hospices and hospitals of the Western world, which, however, did not primarily serve to care for the terminally ill or dying.[4] Research has shown that from 1842 onwards women-directed communities of care gave rise to early in patient institutions in Lyon, Dublin, and London, which already served to care for the dying first and

[1] Shirley Du Boulay, *Cicely Saunders. Ein Leben für Sterbende*, Innsbruck 1987.

[2] Balfour M. Mount, "The Royal Victoria Hospital Palliative Care Service. A Canadian experience", in: *Hospice care on the international scene*, ed. by Cicely Saunders and Robert Kastenbaum, New York 1997, pp. 73–85.

[3] Historical survey in Déon Cox Hayley and Greg A. Sachs, "A brief history and lessons learned from twin efforts to transform medicine", in: *Clinics in geriatric medicine* 21 (2005), pp. 3–15 (with a chronological list of major developments from 1967 onwards); Paul Schotsmans and Tom Meulenbergss (eds), *Euthanasia and palliative care in the Low Countries*, Leuven, Paris, and Dudley, MA, 2005; D. Zech, "Entwicklung der Palliativmedizin in Deutschland", in: *Palliativmedizin heute*, ed. by E. Klaschik and F. Nauck, Berlin, Heidelberg, and New York 1994, pp. 85–102; Helmut R. Zielinski, *Palliative Therapie und Hospizbewegung in der Bundesrepublik Deutschland*, Saarbrücken 1993; Patrice van Eersel, *Réapprivoiser la mort. Nouvelles recherches sur l'expérience de mort imminente*, Paris 1997; C. Centeno and K. S. Heller, "Palliative care in Spain: an evolving model", in: *Journal of palliative medicine* 3 (2000), pp. 123–127; F. Larue, A. Fontaine, et alii, "France: status of cancer pain and palliative care", in: *Journal of pain and symptom management* 12 (1996), pp. 106–108; Anneli Vainio, "Palliative care in Finland", in: *Palliative Medicine* 4 (1990), pp. 225–227; J. D. Cavenagh and F. W. Gunz, "Palliative hospice care in Australia", in: *Palliative medicine* 2 (1988), pp. 51–58.

[4] Jésus Conde Herranz, "Les soins palliatifs: les origines, les antécédents et l'histoire vus à partir d'une perspective chrétienne", in: *Dolentium hominum* 20 (2005), pp. 54–63.

foremost.[5] With good reason, Martha Stoddard-Holmes, Nic Hughes, and David Clark have furthermore drawn attention to the fact that medical doctors in the England of the late nineteenth century were increasingly interested in questions of palliative care.[6] Yet, even the most recent publications often still date the emergence of palliative care in a stricter sense to the period after the Second World War, sometimes even to the exact year of 1967,[7] or else to 1973, the year Balfour Mount supposedly coined the term "palliative medicine".[8]

So far, little thought has been given to the idea that palliative medicine as a form of care thoroughly considered and implemented by physicians might have a much longer history.[9] Some authors are instead outspoken about the reasons why, purportedly, the necessity of hospices and special palliative care did not even arise as a subject of discussion prior to the nineteenth century. As Nicolaus Eschenbruch argues, "Only with modern medicine has it become [...] possible to circumscribe 'dying' as an extended phase of life and thus define it as a subject of hospice work". He holds that, in the past, dying was "either brief, scarcely predictable and brutal or else [meant] a long time of infirmity that was considered God-given and simply part of life". According to him, it was only thanks to new

[5] Clare Humphreys, "'Waiting for the last summons'. The establishment of the first hospices in England 1878–1914", in: *Mortality* 6 (2001), pp. 146–166; eadem, "'*Undying spirits*': *religion, medicine and institutionalized care of the dying 1878–1938*". PhD thesis, Sheffield 1999; David Clark and J. Seymour, *Reflections on palliative care*, Buckingham 1999; E. Klaschik and F. Nauck, "Historische Entwicklung der Palliativmedizin", in: *Zentralblatt für Chirurgie* 123 (1998), pp. 620–623; Grace Goldin, "A protohospice at the turn of the century. St. Luke's House, London, from 1893 to 1921", in: *Journal of the history of medicine and allied sciences* 3 (1981), pp. 383–413.

[6] M. Stoddard Holmes, "'The grandest badge of his art': three Victorian doctors, pain relief, and the art of medicine", in: *Opioids and pain relief. A historical perspective*, ed. by Marcia L. Meldrum, Seattle 2003, pp. 21–34; Nic Hughes and David Clark, "'A thoughtful and experienced physician': William Munk and the care of the dying in late Victorian England", in: *Journal of palliative care* 5 (2004), pp. 703–710.

[7] See, for example, Susanne Ringskog and Danuta Wassermann, *Hastening the end of life. History, research and current Swedish and international debate on the issue of euthanasia*, Stockholm 2000, p. 86: "The palliative medicine takes its start in London, in 1967.".

[8] J. Andrew Billings, "What is palliative care", in: *Journal of palliative medicine* 1 (1998), pp. 73–81, cit p. 73.

[9] J.-M. Nuñez-Olarte, "Care of the dying in eighteenth century Spain – the non-hospice tradition", in: *European journal of palliative care* 6 (1999), pp. 23–27.

therapeutic possibilities that dying "changed from something omnipresent and matter-of-course to a more closely and more clearly defined phase of life".[10] This claim overlooks that in centuries past, countless people died of cancer, consumption, "dropsy", and other chronic illnesses after lying sick for months and suffering the most agonizing pain, and that many of them desperately tried to find help for their pain and other complaints that tormented them. Indeed, the general population knew not only of various ways to relieve the suffering of terminally ill people but even how to hasten death using more or less drastic means.[11]

It is furthermore hardly helpful when some authors, writing of the "death of pain", ascribe a key role to the discovery of anesthesia in the middle of the nineteenth century. They hold that anesthesia fundamentally changed the significance of pain: it was no longer accepted as a necessary evil or divine trial and only now, they claim, were people attempting to make palliative pain relief possible.[12] How this thesis may be reconciled with the centuries-old tradition of treating patients, especially the dying, for their pain using opium, poison hemlock, and other medical pain remedies cannot be said.

The history of the palliative treatment of the incurably ill and dying goes back much further in fact. In the following I will show that the *cura palliativa,* as it was often called, was already delineated and discussed by physicians in the early modern period, including with regard to its range of application and its practical challenges. I will further sketch out the earlier and comparatively broader understanding of what "palliative" treatment meant, along with the negative connotations of certain forms of a "palliative cure" that were sometimes brought into the discussion—connotations that, in some respects, are still part of the discussion of palliative medicine today. As these developments and debates remain largely unknown today, this article is breaking new ground. For this reason, I will limit myself to giving an initial overview, one that is relatively general and essentially focuses on the history of the term. I will work to distinguish and evaluate the different varieties of the *cura palliativa* in the

[10] Nicholas Eschenbruch, "Ein besseres Sterben? Die Entstehung der modernen Hospizbewegung und ihre historischen Voraussetzungen", in: *Praxis* 93 (2004), pp. 1265–1267.

[11] Michael Stolberg, "Active euthanasia in early modern society. Learned debates and popular practices", in: *Social history of medicine* 20 (2007), pp. 205–221.

[12] Stoddard Holmes, Three Victorian doctors [see note6], pp. 21–24.

context of the contemporary therapeutic ideals, as well as outline its relation to the so-called *euthanasia medica,* that is medically assisted dying in the narrower sense. I will leave for future publications a more detailed description of palliative care as practiced by early modern physicians, a description of the medications and procedures that were used, as well as an analysis of national and confessional differences.[13]

History of the Terms Paregorica and Palliativa

In the early modern Christian West, the work of the physician was generally seen to consist not only in healing people who had fallen ill, but also in assisting incurably ill and dying patients. Franciscus Sylvius (1614–1672), for example, stated that quite often a patient's symptoms were so grave that the physician had to disregard the disease itself and fight the symptoms—for instance in cases where a massive loss of blood or other bodily fluids or else serious pain exhausted the vital spirits, or in cases where massive flatulence and winds hindered breathing.[14] On the one hand, the Hippocratic dictum that physicians must not treat incurably ill patients was certainly known and quoted often.[15] But it was also known that this rule was contradicted by other passages in the Hippocratic writings that unambiguously deemed the treatment of incurably ill patients to be the task of the physician.[16] Guido Guidi (early 16th cent. to 1569) concluded that even Hippocrates himself had called for the treatment of incurably ill patients.[17]

[13] This is [2024: was] a major aim of the research project "Geschichte der Palliativmedizin" at the Institut für Geschichte der Medizin at the University of Würzburg, Germany, funded by the Deutsche Forschungsgemeinschaft.

[14] Franciscus Sylvius, "De methodo medendi", in: idem, *Opera medica*, Geneva 1681, pp. 34–62, cit. p. 37.

[15] Hippokrates, "De arte", in: idem, *Œuvres complètes d'Hippocrate*. Ed. by Émile Littré (repr. Amsterdam 1978), Paris 1839–1861, vol. 6, pp. 2–26, cit. pp. 12–14.

[16] Renate Wittern, "Die Unterlassung ärztlicher Hilfeleistung in der griechischen Medizin der klassischen Zeit", in: *Münchener medizinische Wochenschrift* 121 (1979), pp. 731–734; Plinio Prioreschi, "Did the Hippocratic physician treat hopeless cases?", in: *Gesnerus* 49 (1992), pp. 341–350.

[17] Guido Guidi, "De curatione generatim", in: idem, *Opera omnia sive ars medicinalis*, Frankfurt am Main 1626, book 12, cap. 2: "Quaedam mala insanabilia esse et quid requirant.".

Corresponding descriptions of such cases can be found in collections of medical reports from that time. For his part, Sylvius related his practical experience with dying patients at the hospital in Leiden. He wrote that his treatment could no longer do anything to fight the illness of a very weakened patient named Johannes Toeback, who was close to dying. Yet, he was not willing to abandon him to his fate and instead prescribed him fortifying remedies.[18] He also continued to treat a patient with terminal consumption for several weeks, prescribing him remedies that aided expectoration and helped with sleeping. As the man neared his end, he also gave him opiates.[19]

Reports such as these indicate that physicians indeed realized that they needed to change their therapeutic approach when helping these kinds of patients. In general, early modern medicine pursued a causal treatment. The aim was to fight the root source because otherwise the disease would inevitably return. In the early modern period, such a "radical" or "eradicating" treatment—literally taking on the root or tearing it out—usually had the goal of rendering the supposed morbid matter in question harmless and eliminating it from the body. After all, the interpretation of illness as something that resulted from a disturbed balance of the humors and qualities was only of secondary importance in the sixteenth century and later. Early modern physicians for the most part attributed diseases to spoiled, putrefying morbid matter, which could sometimes spread in the form of a contagium. For this reason, almost every therapy was based on bloodletting or cupping, along with a large number of purgatives and emetics as well as diaphoretics. These were complemented with numerous remedies that were believed to have a more or less specific effect on certain diseases and types of morbid matter.[20]

A primarily causal therapy that, in this sense, fought the root of a disease had to be abandoned in the case of incurable and dying patients in favor of a treatment that primarily addressed the symptoms. The necessity of changing the therapeutic approach—today, we would speak about a "change in the therapeutic goal"—was mapped out in a tradition from which early modern authors could draw: the terms taken from the ancient

[18] Franciscus Sylvius, "Collegium nosocomicum", in: idem, *Opera omnia*, Geneva 1698, pp. 709–737, cit. p. 715.

[19] Ibid., pp. 712–713.

[20] Michael Stolberg, *Homo patiens. Krankheits- und Körpererfahrung in der Frühen Neuzeit*, Weimar 2003.

authorities were "paregorein" (παρηγορειν) and "praÿnein" (πραυνειν). Both terms may be translated approximately as "sooth" and "relieve" and are used in a prominent passage in Galen's commentary on the Hippocratic aphorism 6.38, which was quoted often by early modern authors. Here, Hippocrates advised the physician to leave alone (μη θεραπευειν) hidden cancerous tumors (κρυπτοι καρκινοι) that were not yet ulcerated, because treating these forms of cancer risked sending the patient to his grave earlier than if a therapy were omitted.[21] In his commentary, Galen subsequently distinguished two types of therapy. One was the attempt to do everything to make the suffering body part healthy again by removing the disease, including its causes. The other limited itself to alleviating and soothing suffering (παρηγορειν και πραυνειν).[22] Similarly, Dioscorides had previously described some remedies as having a "paregoric" effect,[23] and Aretaios had instructed his readers to use a "paregoric" treatment with patients who suffered from incurable kidney stones and kidney tumors (παρεγορησεται) if necessary.[24]

The ancient poets and historians also used the word *paregorein* to refer to verbal soothing or consoling.[25] When, in the *Deipnosophistae*, Athenaeus writes about the wounded Erypylos, however, we find the specific mention of warm water which, according to Homer, would help relieve his pain (εις δε το παρεγορησαι τας οδυνας), whereas cold water would have been better suited if one merely wanted to staunch the blood,

[21] Hippokrates, "Aphorisms", in: idem, Œuvres [see note 15], vol. 4, pp. 458–609, cit p. 572.

[22] Galen, *Opera omnia*. Ed. by C. G. Kühn, Leipzig 1822 (repr. Hildesheim 1964), vol. 18, pp. 59–61.

[23] For example, Pedanius Dioscorides, *De materia medica libri quinque*. Ed. by Curt Sprengel, Leipzig 1829, vol. 1, p. 37 (Περι αγαλλοχου). In another passage Galen distinguished more precisely "paregoric medicines" (παρεγορικα φαρμακα), which alleviated the symptoms as well as counteracting the disease and πραυνοντα, which only fought the pain and had no effect on the disease as such (Galen [see note 22], vol. 13, p. 707).

[24] Aretaios, *Aretaei Cappadocis opera omnia*. Ed. by Karl Gottlob Kühn, Leipzig 1828, p. 331 (on chronic diseases, book 2, ch. 3).

[25] Plutarch: *Plutarch's lives*. With an English translation by Bernadotte Perrin. vol. 5: *Agesilaus and Pompey. Pelopidas and Marcellus*, London and Cambridge, MA, 1961, p. 144: πραυνεινκαι παργορειν; idem, *Moralia*, vol. VII. Engl. transl. by Philipp H. de Lacy and Benedict Einarson, Cambridge, MA, and London 1959, p. 138 (543A).

as it says later.[26] Above all, it was the term *paregorika* that subsequently became firmly established in medical language. Medical authors writing in late antiquity used it,[27] and early modern physicians were familiar with it as well. It can be found, somewhat Latinized as *paregorica* or in vernacular equivalents such as *parégorique,* in medical dictionaries and encyclopedias dating from the sixteenth to far into the nineteenth century.[28]

Beginning in the late sixteenth century, another related and today still familiar term gained importance, which I will focus on in the following: *palliativa*. It has yet to be verified when the term was coined, and hence we do not know whether it originated in the medical context or was transferred to medicine from other areas of application. The earliest occurrence that I have been able to find is from the second half of the fourteenth century. In an introductory chapter of his *Chirurgia* (ca. 1363), Guy de Chauliac demanded that patients, on principle, be given a specific, causal treatment for their diseases. He then named three situations in which the physician would be allowed to make an exception and limit himself to a *cura larga, praeservativa, et palliativa*: firstly, with diseases such as leprosy, which were fundamentally incurable; secondly, if the patient refused a possible causal, curative treatment or did not listen to the physician's directives; and thirdly, if the curative treatment stood to cause greater harm than the disease itself, for example when the physician was able to staunch bothersome bleedings from hemorrhoids but the body

[26] Athenaeus, *The deipnosophists*, books I-III, 106e. Engl. transl. by Charles Burton Gluck, Cambridge, MA, and London 1951, p. 178.

[27] Oribasius, for example, described the addition of "paregorica" to medicines for patients with massive pain as a situation in which the physician sometimes could not make do with simple medicines (Oribasius, *Œuvres d'Oribase*. Ed. and transl. by Charles Daremberg and Cats Bussemaker. Paris 1851–1876, vol. 2, pp. 741–742).

[28] Bartholomaeus Castellus, *Lexicon medicum graecolatinum [...] ex Hippocrate, et Galeno desumptum*, Messanae 1598, p. 307; Emmanuel Stupanus, *Lexicon medicum graeco-latinum compendiosiss[imum] a Bartholomaeo Castello Messanense inchoatum*, Basel 1628, p. 263; Adrianus Ravesteinus, *Lexicon medicum graeco-latinum a Bartholomaeo Castello Messanense inchoatum*, Rotterdam 1651, pp. 371–372; Stephanus Blancardus, *Lexicon novum medicum graeco-latinum*, Leiden 1690, p. 475; ibid., 5th edn. Halle and Magdeburg 1718, p. 246; Joh. Cunradus Dieterichius, *Iatreum hippocraticum*, Ulm 1661; Louis-Jacques Bégin et alii, *Dictionnaire des termes de médecine, chirurgie, art vétérinaire, pharmacie, histoire naturelle, botanique, physique, chimie*, Paris 1823, p. 446 ("parégorique"); *Dictionnaire des sciences médicales*, vol. 39, Paris 1819, p. 285 ("parégorique", "paregoricus").

required regular bleeding to rid itself of pathogenic substances.[29] In an English translation of Guy's work, published in 1543, we also find the oldest known vernacular use, which contrasts the *eradicatyue cure* with the *palliatyue cure*.[30]

It can be determined with some accuracy, on the other hand, when terms like *palliatio* and *palliativ* came into wider use among physicians in the early modern period. Decisive in this respect was the expansive collection of medical case histories, or *observationes*, penned by the Dutch physician Pieter van Foreest (1522–1597) in the late sixteenth century.[31] In several places, Foreest clearly describes "palliative" therapy. He tells the story, for example, of a seventy-year-old woman from Delft whom he treated in 1560 for a breast tumor.[32] He writes that he declared the woman's condition incurable and inoperable. After giving a mild purgative, he had someone apply medication to the breast "for palliation, as we say" (*ad palliationem ut vocant*). The tumor did not become larger and the woman lived for quite some time according to the description. Writing about other cases, Foreest even asserted that a palliative procedure was superior to more drastic, supposedly radical measures. For example, he advised a woman with breast cancer not to take any medication that would promote an ulceration of the tumor—it was widely hoped at the time that the body would be able to free itself from the cancerous morbid matter by shedding the secretions of the tumor. Instead, he recommended purgatives and bloodletting, apparently for the purpose of carefully draining

[29] Guy de Chauliac, *Chirurgia*, Leiden1559, a2verso-a3verso; cf. idem, *Guigonis de Caulhiaco inventarium sive chirurgia magna*. Ed. by Michael R. McVaugh, vol. 1: *Text*, Leiden 1997; the manuscript on which Mc Vaugh based his edition offers a virtually identical wording and was compiled around 1373.

[30] Guy de Chauliac, *The most excellent workes of chirurgerye*. Transl. by Jean Vigon, London1543, fol. 43v: "We wyll speake of his cure aswel eradicatyue as palliatyue.".

[31] Pieter van Foreest, *Observationum et curationum medicinalium libri XXXII*, Leiden 1603–1606; on the gerne of *observationes*, which was very popular in the early modern period, see Pedro Laìn Entralgo, *La historica clinica. Historia y teoria del relato patografico*, Barcelona 1961; Michael Stolberg, "Formen und Funktionen ärztlicher Fallberichte in der Frühen Neuzeit (1500–1800)", in: *Fallstudien: Theorie – Geschichte – Methode*, ed. by Johannes Süßmann, Susanne Scholz, and Gisela Engel, Berlin 2007, pp. 81–95 [Engl. translation in this volume].

[32] Foreest [see note 31], Leiden 1603 (book 16, "De pectoris pulmonisque vitiis ac morbis" and book 17, "De cordis ac quibusdam mammillarum affectibus", first publ. 1593), pp. 482–486.

the body of the cancerous matter without provoking the cancer. Afterward, she was to use external remedies for a "palliation of the cancer" (*iis topicis uteretur quae ad palleationem cancri facerent*). The woman did not take his advice, however, and went to an *empiricus*, who cut her breast open. An ulcer formed but she kept getting worse. She lived a while longer, suffering dreadful pain, and died, according to Foreest's report, a miserable death.[33] Completely in line with the old Hippocratic idea that cancers should not be irritated unnecessarily, Foreest also advised a third woman with non-ulcerated breast cancer not to provoke the cancer. But she too saw an *empirica*, who succeeded in making the breast fester. Then the patient's arms became swollen, Foreest wrote. She gave off such a stench that it was hardly possible to be in her presence. She became remorseful and returned to Foreest's care, but it was too late: all he could do at this point was carry out a temporary palliative treatment (*palliatione ad tempus in ea usi sumus*), using remedies for the pain and the sore skin.[34]

The question as to where Foreest found the term *palliatio* or *palliare* must remain unanswered. He otherwise quite carefully cites the authorities he is drawing from, but here he only explains his terminology in two places, offering an "as it is called" and "as we say" (*ut vocatur* and *ut dicitur*), which may indicate an oral tradition.[35] The physicians of subsequent times in any case considered him to have introduced the term to medicine. At first, it was found predominantly with Dutch physicians. Henricus ab Heer, for example, told of a *cura palliativa* that he had carried out in 1605 with a seventy-year-old woman who suffered from cancer of the womb. The term *palliatio* was increasingly adopted in other countries as well. Paolo Zacchia (1585–1659), active in Rome, discussed the treatment "that is called palliative" (*quae palliativa dicitur*) as a third fundamental form of treating illness besides the preventative and the curative. He held that it was the physician's human and Christian duty to slow the progression of the disease and mitigate the patient's agony when the disease was incurable.[36] In Hamburg, the Portuguese physician Roderigo da Castro (ca. 1550 – 1627) recommended the use of *paleativa* (sic)

[33] Ibid., p. 485.

[34] Ibid.

[35] Blancardus, Lexicon [see note 28], p. 468, "palliatio".

[36] Paolo Zacchia, *Quaestiones medico-legales*, 3rd edn, Amsterdam 1651, pp. 392–393.

when a case of breast cancer had progressed too far to be treatable with surgery.[37]

As early as 1656, Johannes Antonides van der Linden (1609 – 1664) described the *medicatio palliativa* in general terms for his readers. He called it an important form of therapy, saying that it did not aim to eliminate the disease itself but to besoothe and becalm it. In the case of pain, this mitigation was to be done with anesthesia, while in the case of bleeding and other fluxes, the pathways toward the outside were to be closed. Tumors, on the other hand, were to be dried out so that they would form a scab.[38] Sylvius even gave detailed practical instructions for the *palliatio* of different illnesses. For example, in cases of severe consumption—likely the most common of the generally fatal chronic diseases at that time—he advised the use of hyssop and other remedies to dilute the mucus and thus ease expectoration. And he stated that suitable medicines had to be used to remove the acridity of caustic slime. For loss of appetite, he recommended an elixir that contained ingredients such as myrrh, crocus, aloe, and sulfuric oil. This remedy, he claimed, also supported digestion, that is the fermentation of food, and thus it eliminated the common unpleasant feeling of fullness in the belly after eating. He summarized his recommendations, saying that these few, wisely used remedies helped give fresh courage to consumptive patients, even in this time when they were beyond the hope of complete cure.[39] At the German protestant universities[40] with their thriving culture of disputation and dissertation, the palliative treatment of illness first began to appear as the subject of monographic publications in the late seventeenth century. Elias Küchler's dissertation *De cura palliativa* is the earliest known monograph on the theory and practice of palliative medicine; it was published in Erfurt in 1692.[41] Another work of the same title followed in Jena in

[37] Roderigo da Castro, *De universa muliebrium morborum medicina*, 4th edn, Hamburg 1662, pp. 139–142.

[38] Johannes Antonides van der Linden, *Selecta medica et ad ea exercitationes batavae*, Leiden 1656, pp. 463–466.

[39] Franciscus Sylvius, "Praxeos medicae idea nova", in: idem, *Opera omnia*, Utrecht and Amsterdam 1695, pp. 149–872, cit. pp. 701–702.

[40] I have not found similar monographic studies from other countries.

[41] Elias Küchler, *De cura palliativa*. Praes. Henricus Christophorus Alberti, Erfurt 1692; at the time, medical dissertations were often written by the professor or *praeses*;

1703, written by Georg Wolffgang Wedel (1645–1721).[42] In 1742, again in Erfurt, Georg Ludwig Rosa earned his doctoral title with a thesis on the *Curatio palliativa*.[43] All three authors emphatically urged their colleagues to take the palliative treatment of diseases seriously and consider it an important medical task. They described the areas of use for a *Palliativkur* in great detail and offered references for its practical application with specific diseases. As illnesses that often could only be palliated, Rosa named those diseases where the sheer mass of morbid matter rendered its complete removal impossible—as with large kidney stones or serious asthma that produced large deposits of phlegm in the lungs—but also diseases whose roots had become too deeply established in the body, especially cancer, chronic scurvy, syphilis, epilepsy, melancholy, leprosy, and cachexia.[44] Beyond this, a physician sometimes had to choose a palliative treatment because the overall health of the patient did not allow for a causal therapy. Using the example of torturous anal fistulas, Foreest had already explained that it was sometimes necessary to palliate (*palliare*) because a patient was perhaps too weak to undergo surgery or one had to worry about the patient's postoperative deterioration, for example, if there were a constant discharge of feces.[45] Rosa also added that some patients did not tolerate the necessary medication.[46]

By the second half of the seventeenth century, the term *palliativ* had already found its way into medical dictionaries. Established by Bartolomeo Castelli (d. 1607), the leading *Lexikon medicum graecolatinum*, which went through countless revised editions, was at first content to keep using conventional terms like *Paregorica,Anodyna* (pain

since the author, in this case, repeatedly refers to a professor in Erfurt as his teacher ("praeceptor meus"), the respondant, Elias Küchler, probably wrote the dissertation himself, however.

42 Georg Wolffgang Wedel, *De cura palliativa*. Subm. Philipp Friedrich Schmeltzer, Jena 1703; the frontispiece carries the erroneous year 1603 as the date of publication. It is not clear, who wrote the dissertation but the wide-ranging approach suggests that it was Wedel rather than his student Schmeltzer.

43 Georg Ludwig Rosa, *De curatione palliativa*. Praes. Andreas Elias Büchner, Erfurt 1742; in this case Rosa explicitly signed as the author.

44 Ibid., p. 42.

45 Pieter van Foreest, *Observationum et curationum chirurgicarum libri quatuor posteriores, de vulneribus, ulceribus, fracturis, luxationibus*, Leiden 1601, pp. 333–343.

46 Rosa, De curatione [see note 43], pp. 36–40.

medication) and *Mitigatoria* ("alleviating remedies").[47] But toward the close of the century, Jakob Pankraz Bruno (1629–1709) included *Palliatio* as a separate headword. His definition is avowedly similar to that of Foreest: "Palliatio, or palliativa cura is what the physicians call it when, with desperate and incurable diseases and following an infaust prognosis, certain remedies are given to ease the pain or other burdensome symptoms, as with cancerous ulcers, cancerous fistulas, and others".[48]

An almost literal English translation was offered in the eighteenth century in Robert James's widespread *Medicinal dictionary:* "when in desperate and incurable diseases after predicting the fatal event they prescribe some remedies for mitigating the pain, or some other urgent symptoms, as in ulcerated cancers or cancerous fistulas, and the like".[49] By the nineteenth century, the notion of a "palliative treatment" had entered general usage; it can be found in countless medical treatises and lexicons.

In the early nineteenth century, at the latest, the term "palliative medicine" was already in use, or rather its French equivalent. The *Encyclopédie méthodique* of 1819 has a separate entry dedicated to "médecine palliative". Its author states that no other field of medicine challenged the medical art and the physician's knowledge quite like it. It ultimately touched on all areas of practical medicine. He points out pain treatment with opium in particular, but smaller surgical interventions as well could serve as palliative measures.[50]

Previously, in the course of the eighteenth century, the terms "palliative" and "palliative cure" had spread beyond the borders of the medical sciences, becoming increasingly established in colloquial language. This is impressively seen in its widespread metaphoric use. Just looking in the

[47] See note 28.

[48] Jacobus Pancratius Bruno, *Castellus renovatus, hoc est, Lexicon medicum, quondam a Barth. Castello Messanensi inchoatum*, Nürnberg 1682, pp. 875–876: "Palliatio, palliativa cura, vocatur a medicis, quando in morbis desperatis et incurabilibus, praemisso prognostico eventus funesti, quaedam remedia mitigantia vel dolorem, vel alia symptomata urgentia adhibentur, ut, in cancris ulceratis, fistulis cancrosis, aliisque." In his *Lexicon*, Blancardus [see note 28], p. 475, listed "plaats-middelen" and "streel middelen" as Dutch synonyms. The latter was apparently derived from the Dutch term "strelen" for "to alleviate" or "to carress".

[49] Robert James, *A medicinal dictionary, including physic, surgery, anatomy, chymistry, and botany*, 3 vols, London 1743–1745, entry on "palliatio".

[50] *Dictionnaire des sciences médicales*, vol. 39, Paris 1819, p. 123; *Encyclopédie méthodique. Médecine*, vol. NOY-PHT, Paris 1824, pp. 283–284.

voluminous Krünitz encyclopedia, there are around fifty entries that use terms like "palliative remedy" or "palliative cure", some of them in a medical context and some in moral or economic contexts.[51] For example, the temporary buying-up of crops was considered an "excellent palliative remedy" to stop the deterioration of crop prices,[52] and Johann Wolfgang von Goethe declared "Liebeley" or flirting with the other sex the "most effective palliation" for his "devilish mood".[53]

The Palliative Cure and Euthanasia Medica

Quite apart from such metaphoric uses, the understanding of "palliative" treatment at that time was considerably broader than today, however, even in the medical context. One important difference concerns the point in the course of a disease or a life at which a "palliative" treatment seemed indicated. Today, the notion of "palliative medicine" is applied almost exclusively or at least primarily to terminal patients who are close to dying, even though the oft-quoted definition of the WHO goes beyond that, describing "palliative" treatment options for the earlier stages of incurable, usually fatal diseases.[54] The pre-modern "palliative cure", by contrast, was generally indicated for chronic incurable diseases accompanied by irritating or agonizing symptoms that, while not the cause of death, would quite possibly not go away until the patient had died.

In this sense, easing the suffering of the dying, the *cura mortis palliativa,* as it was called in the title of a medical dissertation in 1723,[55] was only a subdomain at the time. The author also wrote about *euthanasia palliativa,* which the dying themselves or those present at the sickbed

[51] Johann Georg Krünitz, *Ökonomisch-technologische Encyklopädie oder allgemeines System der Staats-, Stadt-, Haus- und Landwirtschaft, wie auch der Erdbeschreibung, Kunst- und Naturgeschichte in alphabetischer Ordnung*, Berlin 1773–1858.

[52] Ibid., vol. 45 (1789), p. 629.

[53] Johann Wolfgang von Goethe, *Goethes Briefe an Frau von Stein.* Ed. by Jonas Fränkel, vol. 1. Jena 1908, p. 2 (letter from Goethe, January 27, 1776).

[54] Editors' "Introduction", in: *Oxford textbook of palliative medicine*, ed. by Derek Doyle, Geoffrey Hanks, and Neil MacDonald, Oxford 1998, pp. 3–8; Geoffrey P. Dunn and Robert A Milch, "Introduction and historical background of palliative care: where does the surgeon fit in?", in: *Journal of the American College of Surgeons* 193 (2001), pp. 325–328.

[55] Georg Christoph Detharding, *De mortis cura.* Praes. Georg Detharding, Rostock 1723; it remains unclear whether the father, the son, or both wrote this dissertation.

requested so "they may die gently", as people said, "without having a hard end [and instead] going to sleep".[56] In the eighteenth and nineteenth centuries, as the term *euthanasia* or *euthanasia medica* came into usage, the medical treatment of the dying developed into an independent subject and area of interest. Under Michael Alberti (1682 – 1757) in Halle, a doctoral dissertation titled *De euthanasia medica, Vom leichten Todt* (De euthanasia medica, On Easy Death) was defended in 1735.[57] Around 1800, Nicolaas Paradys (1740 – 1812) in Leiden garnered great attention far beyond the borders of the Netherlands with his *Oratio de euthanasia naturali*. The speech was translated to German just two years later.[58] Approximately two dozen dissertations were written at German, Dutch, and Swedish universities in the course of the subsequent decades, discussing specifically *euthanasia medica*[59] or else something called the "euthanasic curative method".[60] Leading contemporary physicians also explored the subject in works on general medicine.[61]

Denoting a good, a gentle death, the word *euthanasia* had been passed down from antiquity. In religious devotional writing, it came to be used in an active sense at the latest in the sixteenth century, now referring to a person's spiritual preparation for a good death that would be pleasing to God.[62] In the early seventeenth century, Francis Bacon (1561–1626)

[56] Ibid., p. 85.

[57] Michael Alberti, *De euthanasia medica, Vom Leichten Todt*. Resp. Zacharias Philippus Schulz, Halle and Magdeburg 1735.

[58] Nicolaas Paradys, *Oratio de euthanasia naturali et quid ad eam conciliandam medicina valeat*, Leiden 1794; idem, "Rede von Nikolaus Paradys, Professor in Leyden, über das, was die Arzneiwissenschaft vermag, den Tod leicht und schmerzlos zu machen", in: *Neues Magazin für Aerzte* 18 (1796), pp. 560–572.

[59] See, for example, Friedrich Kessler, *De euthanasia medica sive de moribundorum adjuvandorum ratione*. Diss. med., Berlin 1828; Julius Stubendorff, *De euthanasia medica*, Dorpat 1836; Wilhelm Baltes, *De euthanasia dissertatio inauguralis medica*, Berlin 1842; Heinrich Brockerhoff, *De euthanasia medica*, Berlin 1843; cf. Christoph Hoffmann, *Der Inhalt des Begriffes "Euthanasie" im 19. Jahrhundert und seine Wandlung in der Zeit bis 1920*, Berlin 1969.

[60] W. Hennemann, "Kakothanasie", in: *Beiträge Mecklenburgischer Ärzte zur Medicin und Chirurgie* 1 (1830), pp. 175-80, cit. p. 177.

[61] See, for example the chapter "Euthanasia, oder von den Hülfen erträglich zu sterben" in J. C. Reil, *Entwurf einer Allgemeinen Therapie*. Ed. by P. Krukenberg, Halle 1816, pp. 560–582.

[62] Martin Eisengrein, *Euthanasia sive, de firma spe ac fiducia in Dei misericordia, mortis tempore, collocanda assertio catholica*, Cologne 1576.

expressly asked physicians to resort to *euthanasia* in their encounters with the dying. They were to relieve dying patients' complaints and help them have a gentle death.[63] It was to this tradition of medical care for the terminally ill that the physicians practicing in the eighteenth and nineteenth centuries avowedly referred—they did not have in mind any life-shortening measures in the modern sense of "active euthanasia".

Some publications about *euthanasia medica* explicitly referred to medical treatment that was deemed appropriate for the dying as "palliative".[64] In great detail, they describe the possible applications and dangers of poison hemlock, henbane, laurel water, thornapple, musk, camphor, and other analgesic and antispasmodic or else restorative or stimulating remedies. They considered opium in particular to be a blessing, though of course it had to be carefully dosed.[65] At the same time, the necessity of deliberately forgoing therapy was underscored. There were urgent admonitions against trying against all odds to turn hopeless cases around. The risk authors saw was that, instead of a good death, a *Euthanasia*, one might cause a bad death, a *dysthanasia* or *kakothanasia*.[66] Apparently showing this kind of reserve could not be taken for granted. Wilhelm Hennemann (1786–1843), for example, complained bitterly about the "reprehensible medical drive" of colleagues who left "no remedy untried, not the cruelest, not the most abhorrent, to save the sick, [...] even when the hope for success lacks any inherent probability".[67]

Euthanasia medica was not restricted to medical treatment in a narrow sense. Nursing aspects played a major role: for example, making sure the air was clean, preventing pressure ulcers by providing clean bedding and repositioning the patient frequently, choosing the right nutrition, and so

[63] Francis Bacon, *De dignitate et augmentis scientiarum libri IX*, Paris 1624, pp. 222–223; on the history of the term "euthanasia" see Udo Benzenhöfer, *Der gute Tod? Euthanasie und Sterbehilfe in Geschichte und Gegenwart*, Munich 1999.

[64] Kessler, De euthanasia [see note 59], p. 7.

[65] Wilhelm Ludwig Ziemssen, *De euthanasia medica*. Diss. med., Greifswald 1845, pp. 45–46.

[66] Michael Alberti, *De dysthanasia medica. Vom schwehren Todt*. Resp. Karl Christian Hennig. Halle and Magdeburg 1735; Jakob Dietrich Ludwig Jahn, *De euthanasia dissertatio inauguralis medica*, Kiel 1839, p. 10.

[67] Hennemann, Kakothanasie [see note 60], p. 175.

on. Special attention was also paid to the *cura psychica*,[68] the psychological wellbeing of the dying and their relatives. In its explanation of *Médecine palliative*, the *Encyclopédie méthodique* had already pointed out that the most powerful "palliatives were those that worked on the soul, that belonged to the art of caring for the soul ("cet art de traiter les ames").[69] In this sense, some writings about *euthanasia medica* put an emphasis on the beneficial effect of good (though limited) company, gentle music as well as fragrant flowers, paintings, and sculptures. The concern with the psychological wellbeing of the dying also led to the central question of how truthful one should be at the sickbed.[70] The general consensus was that the patient must not be told about their bad prognosis, unless imperative worldly matters such as the indispensable completion of a last will and testament were a compelling force. According to dominant doctrine, this was because the shocking news threatened not only to affect the patient's mood but also to have a negative effect on the physical disease process itself, and thus it threatened to hasten death.[71] Therefore, only the relatives were supposed to know. Some authors even preferred to leave the task of comforting the soul to the physician rather than the cleric. The physician would be perceived as a source of hope and, because of his knowledge of the patient and the patient's temperament, he could adapt his words of support to the patient's condition and mood. If, however, the cleric were called, this would give the patient a terrible sense of his imminent death. Not only this, but some physicians complained that clerics sometimes even painted

[68] Richardus Heinzelmann, *De euthanasia medica*. Diss. med., Berlin 1845, p. 22.

[69] *Encyclopédie méthodique. Médecine*, vol. NOY-PHT, Paris 1824, pp. 283–284; on contemporary.

"médecine morale", see Elizabeth A. Williams, *The physical and the moral. Anthropology, physiology, and philosophical medicine in France, 1750–1850*, Cambridge 1994.

[70] See also Karen Nolte, "Zeitalter des ärztlichen Paternalismus? Überlegungen zu Aufklärung und Einwilligung von Patienten im 19. Jahrhundert", in: *Medizin, Geschichte und Gesellschaft* 25 (2006), pp. 59–89.

[71] [See Michael Stolberg, "Emotions and the body in early modern medicine", in: *Emotion review* 11 (2019), pp. 113–122; in the original publication of this article a forthcoming contribution to a collective volume was quoted which unfortunately never made it into print].

a vivid picture of the torments of hell for the dying, thus hastening their death.[72]

Palliative Treatment and Curative Ambitions

According to the view unanimously shared by pre-modern medical literature, palliative treatment made sense and was necessary with many incurably ill and dying patients. No reasonable person could doubt that in some cases even the best possible practice of the medical art could do nothing. When the body had become too weak or else the disease or the morbid matter had become too deeply rooted, no therapy was able to eradicate the disease. Here, treatment had to adopt a course of action that was indicated by the symptoms.[73] And yet, the pre-modern "palliative cure" remained in a tension with the dominant therapeutic ideal of causal disease treatment that got to the roots of the disease. After all, palliative treatment, as per its definition, was not causal treatment. In the words of Johann Jakob Woyt (1671–1709), it was an "interim cure" that was applied "with desperate and incurable diseases", when it was used "for example to ease pain, so as not to leave the patient completely helpless, although one achieved little with the cure".[74] Or, as defined in Krünitz's encyclopedia, it was "a course of treatment that does not heal the affliction but only alleviates it for some time".[75] While some authors pointed to the successful mitigation of intense pain or massive loss of blood and other bodily fluids and posited that it had an indirect curative effect because it involved countering something that weakened the body's

[72] Baltes, Euthanasia [see note 59], pp. 18–20; on the tensions between bodily care ("Krankenpflege") and spiritual care ("Seelenpflege") see also Karen Nolte, "Vom Umgang mit Tod und Sterben in der klinischen und häuslichen Krankenpflege des 19. Jahrhunderts", in: *Pflege - Räume, Macht und Alltag. Beiträge zur Geschichte der Pflege*, ed. by Sabine Braunschweig, Zürich 2006, pp. 165–174.

[73] For a detailed analysis of the differences between a "symptomatic indication" and a therapeutic approach that aimed at the "essence" or "nature" of the disease, see Ferdinand G. Gmelin, *Allgemeine Therapie der Krankheiten des Menschen*, Tübingen 1830, pp. 64–71.

[74] Johann Jakob Woyt, *Gazophylacium Medico-Physicum oder Schatz-Kammer, medicinisch- und natürlicher Dinge*, Leipzig 1709, p. 673.

[75] Krünitz, Encyklopädie [see note 51], vol. 106 (1807), p. 284.

natural healing powers,[76] this indirect causal effectiveness of a palliative cure could hardly be claimed for all relevant cases of illness. Rather, a "merely" palliative treatment was often contrasted with a *cura vera* or *cura legitima,* which pulled the disease up by its roots—a relationship that, in some respects, still marks palliative medicine today.

This tension was considerably greater at the time because early modern physicians, thanks to their schooling in Latin, were more keenly aware of the etymological roots of the word *palliare* than most practitioners of palliative medicine today. The term, as we know, derives from the Latin *pallium* for "cloak". *Palliare,* understood rather literally, thus meant "spreading a cloak over something", that is "cloaking" or "covering up".[77] In the widely known medical dictionary of Johann Jakob Woyt of 1709, *palliatio* is therefore translated as *Bemäntelung* or "cloaking".[78] In his translation of Johann Baptist van Helmont's *Aufgang der Artzney-Kunst* (Rise of the Art of Medicine), Christian Knorr von Rosenroth (1636–1689) even invented the term *Mantel-Kuren* or "cloak-cures", which never caught on, however.[79]

This original meaning of *pallatio,* which would have also been obvious to educated laypeople at the time, tells us why early modern physicians—and this seems somewhat surprising today—applied the term *palliativa* to those remedies and procedures that did nothing more than hide or cover external complaints or blemishes, thus concealing them from the senses of others. This, as Wedel explained, was the real, original meaning of *cura palliativa.*[80] For Küchler and Rosa, this "dissimulative" or "external" palliative cure thus encompassed such things as prosthetics and cosmetic surgery, which served to obscure the loss of a limb, the nose, an eye or another body part. Even eyeglasses, pads used with urinary incontinence

[76] Johannes van Heurne, *Praxis medicinae nova ratio,* Leiden 1590, pp. 343–344; Guy de Chaulica, *Chirurgia magna,* Darmstadt 1976 (reprint of the 1585-edn), pp. 398–400; Wedel, De cura [see note 42]; similarly, idem, *Opiologia,* Jena 1682, p. 157.

[77] In the Netherlands, where the Latin term "palliatio" first entered more common usage, "manteln" oder "menteln" were used as synonyms (see Lorenz Diefenbach, *Glossarium latino-germanicum mediae et infimae aetatis,* Frankfurt 1857 [repr. Darmstadt 1968], fol. 407a.

[78] Woyt, Gazophylacium [see note 74], p. 673.

[79] Johann Baptist van Helmont, *Aufgang der Artzney-Kunst.* Transl. by Christian Knorr von.

Rosenroth, Sulzbach 1683, p. 6.

[80] Wedel, De cura [see note 42], p. 4.

or else cosmetics that concealed blemishes or impurities of the skin were named as examples for “palliative” treatment in this sense.[81]

In most cases, when openly visible blemishes were covered up or repulsive smells were masked, this was a positive thing insofar as it benefitted both the afflicted and those around them. In cases where such procedures served to hide the marks of a dishonoring punishment such as a brand mark or a cut-off nose, they might have had a dubious status. But the word “cloak”, as in to “cloak” something, could also be understood to refer to reprehensible or even fraudulent intentions when attempting to hide or conceal something; this seems to be the rather negative connotation of the verb that prevails today. In fact, in German speaking Europe, the etymological proximity of *pallatio* and “bemänteln” (cloaking) was so apparent to early modern physicians that they also applied the term *cura palliativa* when discussing misguided, even disdained forms of medical treatment, that is cases in which diseases and their symptoms were no more than “cloaked”: the healer tried—or pretended to—grab the disease by its roots but, if at all, he only managed to make the external symptoms go away, while the disease continued to wreak havoc inside the body. Here, the “palliative cure” was synonymous with “sham treatment”, and “palliative remedies” were “sham remedies”.[82]

In this context, Johannes Antonides van der Linden, for example, spoke of a *medicatio simulata* in cases where a patient suffering from stone disease was unable to urinate, and a practitioner went about healing him by pushing the stone from the bladder exit back into the bladder. Almost invariably, urinary retention would soon recur. And he went on to say that the same was true if someone boasted about having healed a fever but in fact only suffocated it, without also cleansing the blood of its impurities. The morbid matter remained in the body and the illness would return at the very next opportunity.[83] As various authors explained, Wedel at great length, a purportedly causal treatment that only “cloaked” the symptoms could even make the disease considerably worse. If a physician contented himself with relieving the cough of a consumptive patient, for

[81] Küchler, De cura [see note 41], p. 21; Wedel, De cura [see note 42], pp. 12–19; cf. Rosa, De curatione [see note 43], p. 41; similarly still Edward Lund, *Palliative medicine and palliative treatment in surgical cases. Address delivered at the meeting held in Manchester, June 30th, 1880*, London and Manchester 1880.

[82] Krünitz, Encyklopädie [see note 51], vol. 141 (1825), p. 723.

[83] Linden, Selecta medica [see note 38], pp. 455–456.

example, he would temporarily stop the bothersome urge to cough and the patient would be happy. However, the sputum would soon accumulate in the lungs. The same applied when ulcers, bleeding hemorrhoids or scabies rashes were healed without eliminating the morbid matter that had previously been excreted through them. The morbid matter remained in the body and made things even worse. It became mixed in with the blood mass or attacked the surrounding tissue, possibly going all the way down into the bones.[84]

Descriptions of failed "palliative cures" like these served to remind soon-to-be physicians to practice great diagnostic and therapeutic care. However, their authors did not accuse their colleagues first and foremost of such misguided treatment that only "cloaked" and perhaps worsened the disease. Instead, they pointed at the patients and at their "uneducated" competitors. Rosa complained that, right from the beginning, the sick often made it impossible for the physician to identify and fight the true cause of an illness. They could not properly remember and came to the physician with preconceived ideas, or else they underestimated the paramount influence of personal dispositions and lifestyle on the development of diseases. Others were garrulous to no end, never got to the point, and did not let the physician get a word in edgewise. When the physician finally did manage to get to the bottom of the disease despite all this, patients would refuse the necessary therapy, if only because they did not like it, or else they would forget or disregard the physician's counsel, not observe his dietetic advice, or even try household remedies.[85]

To a far greater extent, the blame for a failed treatment, one that was unintentionally no more than "palliative" and only "cloaked" the disease, was laid on the "uneducated" competition, the *empirici*, the *impostores*. Their popularity and success were a thorn in the sides of the learned physicians of the period. With consternation, physicians experienced on a regular basis how patients, including many educated patients, invested their trust in these *empirici* and were convinced of their abilities and their positive therapeutic outcomes. And so, writing dozens of treatises and pamphlets, physicians appealed to the public and the authorities, warning about the "murderous" doings of the empirics, Jews, and old women, and demanding stern punishment. Looking back today, the fact that they

[84] Wedel, De cura [see note 42], pp. 24–31.

[85] Rosa, De curatione [see note 43], pp. 36–40.

did not meet with much success in these efforts seems to make sense. To laypeople, it was hardly obvious that the physicians' medical treatment was superior because physicians generally relied on remedies and procedures that were very similar to those used by other healers—primarily purgatives and bloodletting.[86]

Faced with this situation, physicians happily seized the opportunity to write discourses about the "palliative cure" as this allowed them to promote themselves professionally. Foreest, the author of one of the best-known medical invectives against the "fraudulent practices" of countless traditional uroscopists,[87] had framed his collection of case histories mentioned at the beginning as instructive *exempla* of the *empirici's* "ignorance". These *empirici*, incapable of delivering adequate diagnoses and therapies, or even being, as the medical criticism held, fraudulent in their intentions, led their patients to believe that they were curing the disease when they made symptoms disappear, or when they seemingly eliminated morbid matter from the body. They may have even been richly rewarded. Balthasar Timäus von Güldenklee (1601–1667), for example, wrote about an *empiricus* who relieved a noblewoman suffering from an itch that had tormented her for years after her husband had infected her with syphilis. The man was rewarded with 500 gold pieces, and had, in Timäus's account, stolen the physicians' prize. However, the *empiricus* had left the disease cause, the syphilitic *causa morbi* in the body. The disease spread all the more violently and was now out of reach to any medical art. The woman died soon after.[88]

Conclusion

As we have seen, modern palliative medicine can look back on a centuries-old tradition. Building on a limited number of ancient references to the use of *paregorica* and *praÿnonta* for incurable diseases, palliative treatment

[86] Ananius Horer, *Artzney-Teuffel, Oder kurtzer Discurs, Darinn diesem Ertzmörder seine Larve abgezogen*, sine loco 1634.

[87] Pieter van Foreest, *De incerto, fallaci urinarum iudicio*, Leiden 1589; English edition. by James Hart under the telling title, *The arraignment of urines, wherein are set downe the manifold errors and abuses of ignorant urine-monging empirickes, cozening quacksalvers, women-physitians and the like stuffe* (London 1623).

[88] Balthasar Timaeus von Güldenklee, *Casus medicinales praxi triginta sex annorum observati*, Leipzig 1667, pp. 277–278.

was described and discussed as an independent and important form of medical practice as early as the sixteenth and seventeenth centuries. In the eighteenth and nineteenth centuries, its practical application specifically for the dying was elaborated. Today, many of the ways of thinking about palliative care that existed back then, as well as the related critiques and demands, still hold sway and remain central to the discussion, whether this is the appreciation for relief of suffering among the dying as one of the highest medical tasks; the warning against an excessive drive to take medical action; the great attention paid to aspects of nursing care; the effort to provide personal care on an individual basis, combined with a criticism of physicians "who believe their concern is only the disease and not the human being"[89]; and, finally, the consideration that is extended to the needs of family members.

With respect to terminology, this long historical tradition proves to be a burden. As we have seen, the literal meaning of *palliatio* as "cloaking" also led contemporary authors to apply the word with its negative connotation of "hiding" or "concealing" to failed treatments that, without good reason, neglected the root cause of a disease. This suspicion of merely "concealing", of forgoing a causal treatment—perhaps without adequately considering the matter—continues to haunt palliative medicine today and goes hand in hand with the worry that sometimes the therapeutic goal is changed prematurely, thus shortening a patient's life.

Modern palliative physicians have occasionally pointed to this problematic etymological connection between the two meanings of "palliation".[90] D. Morris has even suggested that the term "palliative" be construed on the basis of Indo-European root words such as *pel* for skin and *pelte* for shield so as to connect palliative medicine with imagery that is more positive and creative.[91] This, however, stands in the face of centuries of clear derivation from the word *pallium* and its associations with covering and concealing, as has been mentioned.

It appears downright paradoxical, when protagonists of today's palliative medicine for their part inadvertently reinforce the associative link

[89] Baltes, Euthanasia [see note 59], p. 7.

[90] Julia Lawton, "Contemporary hospice care. The sequestration of the unbounded body and 'dirty dying'", in: *Sociology of health and illness* 20 (1998), pp. 121–143.

[91] D. Morris, "Palliation. Shielding the patient from the assault of symptoms", in: *Academy update. American Academy of Hospice and Palliative Medicine* 7 (1997), pp. 1–11, cit. by Clark and Seymour, Reflections [see note5], pp. 80–81.

between "palliative medicine" and a misguided abandonment of causal therapy. Time and again, one finds assertions in the literature that the medicine that was practiced for centuries was, due to limited therapeutic possibilities, mainly of a palliative nature. According to this claim, it was only around 1900, or even as late as around 1950, that a new approach in which actual disease causes were isolated and combatted replaced the previous one.[92] This thesis completely fails to recognize that, beginning in antiquity, Western physicians always made a decisive effort to carry out a causal treatment, one that addressed the root of a disease—though it goes without saying that their ideas about the causes of illness were markedly different from today's. Moreover, the thesis equates "palliative" with "ineffective". But it should be immediately clear that the central pre-modern treatment methods—bloodletting, drastic purgatives, mercury preparations, and so on—did not have a "palliative" effect and did not attenuate subjective complaints—not in any estimation, past or present. Looking at the early modern writings presented here, we see how, in cases where a curative effect could no longer be expected, readers were advised to forgo such physically burdensome procedures and medications precisely because they were not compatible with palliative treatment.

Modern palliative medicine, incidentally, might have done well to call itself "paregorical medicine". Not only would the ambiguous association of *palliare* and "cloaking" be avoided, but the term, by virtue of its etymological roots, would be associated with words of encouragement and comfort and with a responding to the psychological needs of the sick. These were elements to which physicians prior to 1850—perhaps due to an awareness of their limited therapeutic options—were much more attuned than practitioners of later generations.

[92] H. Pichlmaier, "Palliativmedizin (editorial)", in: *Zentralblatt für Chirurgie* 123 (1998), p. 619; P. D. Wall, "Editorial: 25 volumes of 'Pain'", in: *Pain* 25 (1986), pp. 1–4; Doyle, Hanks, and MacDonald, Introduction [see note 54], p. 3; Derek Doyle, "The provision of palliative care", in: *Oxford textbook of palliative medicine*, ed. by idem, Geoffrey W. C. Hanks, and Neil MacDonald, 2nd edn, Oxford 1998, pp. 41–53, cit. p. 41.

Forms and Strategies of Authorization in Early Modern Medicine

Historians of science have increasingly moved away from the traditional notion that new scientific methods, theories, and practices come to prevail simply because they are "true" or at least superior to previous ones. Science has come to been much more as a form of social, cultural practice that always exists in a close interrelationship with its particular institutional, social, cultural, economic, and political context. Taking a social-constructivist perspective, historians have underscored the influence of economic interests, political loyalties, and ideological preferences on the development of science. Laboratory studies have examined the complex processes involved in the transformation of experimental data

First published in German as "Formen und Strategien der Autorisierung in der frühneuzeitlichen Medizin", in: *Autorität der Form – Autorisierungen – institutionelle Autorität*, ed. by Wulf Oesterreicher, Gerhard Regn, and Winfried Schulze, Münster: LIT-Verlag 2003, pp. 124–152. Translated into American English by Logan Kennedy and Leonhard Unglaub.

M. Stolberg, *Medicine and the Body in Early Modern Europe*, Palgrave Studies in Medieval and Early Modern Medicine,
https://doi.org/10.1007/978-3-031-84556-7_6

into scientific "insights" as well as the influence of communicative structures and instruments in this process.[1] Not least of all, research has considered the manifold, largely rhetorical strategies scientists use to lend themselves authority, to advance themselves, and to assert their findings, interpretations, and practices.

It is my aim in this article to outline the central, prevalent forms and strategies of authorization in early modern medicine along with the associated problems.[2] First, I will focus on the processes of constituting, claiming, and challenging authority as we find them in the scholarly communication among physicians. In a second step, I will open the perspective to consider the individual and collective claim to higher status and recognition that was made as learned physicians contended with the numerous competitors in the pluralist medical market.

The Authorization of Medical Knowledge in Physician Discourse: Ratio, Experientia, and Auctoritates

In order to authorize their personal ideas or positions as true and binding, the writers of medical texts in the sixteenth and seventeenth centuries, when they discussed scholarly matters among themselves relied on the traditional triad of *ratio, experientia*, and *auctoritates*. Between the three, there was some tension, but they also reinforced each other to some extent. For example, the doctrines of the ancient physicians attained their great authority not only because they were based in age-old experience but also because they had withstood the rational examination of many generations of physicians.

In the course of the early modern period, a number of characteristic changes and shifts occurred within this triad. Citing ancient or medieval

[1] Bruno Latour, *Science in action. How to follow scientists and engineers through society*, Cambridge, MA, 1987; Ludmilla Jordanova, "The social construction of medical knowledge", in: *Social history of medicine* 8 (1995), pp. 361–381.

[2] Research on this topic is scant; see, however, Wolfgang Eckart, "'Auctoritas' versus 'Veritas' or: Classical authority and its role for the perception of truth in the work of Daniel Sennert (1572–1637)", in: *Clio medica* 18 (1983), pp. 131–140; Richard Toellner, "Zum Begriff der Autorität in der Medizin der Renaissance", in: *Humanismus und Medizin*, ed. by Schmitz Rudolf and Keil Gundolf, Weinheim 1984, pp. 159–179.

auctoritates remained an important and indispensable form of authorization for collective and individual medical views. In some cases, this bordered on quasi-religious exaltation. Hippocrates, Aristotle, Galen, and Dioscorides, with their "authority or power" were the four "evangelists" of medicine, wrote Christophorus Clauser, for example.[3] Not surprisingly, then, most of the contemporary medical authors seeded their discussions with plentiful references to, and quotations by, ancient writers so as to lend them more weight.

However, the preferences for different authorities changed considerably in the course of the early modern period. Sixteenth-century humanism opposed the traditional primacy of Avicenna's *Canon medicinae* and of "Arabian" medicine in general.[4] Taking its place were most notably the works of Galen, who became the preeminent authority.[5] Yet, the humanist physicians' intensive work with sources had paradoxical effects in the longer term. Through their precise philological efforts, they sought to reconstruct the authoritative original text of the classical writings, which would point the way to the one true medicine. Their more complete and precise knowledge of the classical sources, however, confronted them—much more so than previously—with the many contradictions between the different classical authorities and sometimes even in the work of one and the same author, especially Galen.[6] The resulting pluralism of ideas weakened the legitimacy of the individual authorities and their writings and thus their usefulness for underscoring certain positions. At the same time, the diminished weight of the argument "ex auctoritate" created more leeway for alternative strategies of reasoning and validation.

In the medicine of the sixteenth and seventeenth centuries, *experientia* was given increasingly more weight compared to the *auctoritates*, due in part also to the influence of developments in contemporary natural

[3] Christoph Clauser, *Das die Betrachtung des Menschenn Harns on anderen Bericht vnnütz, vnd wie doch der Harn zuoempfachen vnd zuourteylen am geschicktisten syge, mit anderen in der heyligen Artzny Leeren vnd nit vnnützen Warnungen Dialogus, das ist ein Gespräch*, sine loco [1531], no page numbers.

[4] Jerome J. Bylebyl, "The school of Padua: humanistic medicine in the sixteenth century", in: *Health, medicine and mortality in the sixteenth century*, ed. by Charles Webster, Cambridge 1979, pp. 335–370.

[5] Owsei Temkin, *Galenism: Rise and decline of a medical philosophy*, Ithaca, NY, 1973.

[6] Cf. Toellner, Begriff [see note 2].

philosophy. Not only anatomy but also clinical medicine were increasingly perceived as primarily empirical in nature, that is supported by and drawing from experience. Clinical case observations that were modeled in part on the more "empirical", casuistic Hippocratic writings came to complement learned commentary and theoretical treatises. Starting in the late sixteenth century, the *observationes*, *consilia*, or *consultationes* written by well-known physicians constituted one of the most popular medical genres.[7] Physicians and scholars still readily quoted such case observations decades later to argue or consolidate their own positions.

The growing significance of empirical observations in the practice of physicians at the same time marked a radical and powerful change in the grounding of medical truth claims. The empirical method moved the foundations of medical knowledge into an area beyond human thought and the written or spoken word. Medical knowledge was now finding its basis in something exterior, something that was "objectively" given to human perception, which preceded all human theory. As William Harvey wrote, defending himself against conservative criticism: facts that are obvious to the senses ("facta manifesta sensui") are not "opinions" ("opiniones"). Nature, he held, was much older than the oldest authorities of medical literature and therefore could not be disproven by them.[8]

In practice, however, the validation and authorization of medical knowledge by way of observation and experience proved problematic. How was one to distinguish between correct observations and flawed or even invented or fabricated observations? Adequate criteria were needed. Central at the time was the credibility, the personal authority of the one who had made an observation and shared it. The individual physician had to prove his personal authority accordingly, for example by referring to his many years of experience or to his training with the best physicians and natural philosophers. Beyond this, contemporary physicians relied on a number of rhetorical and representational strategies to lend additional validity to their empirical observations.

[7] The most famous example is Pieter van Foreest, *Observationum et curationum medicinalium sive medicinae theoricae et practicae libri XXXII*, Nürnberg 1676; on the genre, see Pedro Laìn Entralgo, *La historica clinica. Historia y teoria del relato patografico*, Barcelona 1961.

[8] William Harvey, *Exercitationes anatomicae, De motu cordis et sanguinis circulatione*, Rotterdam 1671, p. 242.

In anatomy, one of the most important methods was visualization. Empirical findings could be visualized in great detail thanks to new means of representation that resulted in realistic images with accurate perspective, maybe even with a similarly realistic landscape or city view as a background, which gave the findings they illustrated a compelling degree of evidence. The rhetorical authorizing function of such representations becomes immediately obvious today in cases where structures were meticulously pictured that we no longer know in this form or accept as true today, for example a heart-shaped hole in the female sternum, or embryos in the mother's belly that look like four-year-old toddlers—reflections of prevailing cultural assumptions about gender difference and about the nature of the fetus as complete and ensouled.[9]

Another way of authorizing empirical knowledge, especially in case histories and *observationes*, was to provide the exact place and time when an observation was made. Without considering a person's reputation and "privacy" (which did not exist as such at the time, or at least not in the form it takes today), physicians published their clinical observations with the exact name of the patient and the place and time, thus guaranteeing (or suggesting) the possibility that their claims could be verified.

A third strategy was to cite "credible" witnesses who had made the same observation, ideally while present at the same time. This became the most important new epistemological ideal for those connected with the circle of the Royal Society.[10] While observing experiments together in front of invited witnesses was less relevant in medicine, physicians in their case histories and autopsy reports tended to name several colleagues who had been present during a patient visit or a dissection who could testify to what had been observed.

Finally, the authorizing role of *ratio* as the third element of the triad remained indispensable, despite the manifold criticism leveled at the scholastic tradition. For a medical theory and practice that was based in

[9] Michael Stolberg, "A woman down to her bones. The anatomy of sexual difference in the sixteenth and early seventeenth centuries", in: *Isis* 94 (2003), pp. 274–299; Barbara Duden, "Zwischen 'wahrem Wissen' und Prophetie. Konzeptionen des Ungeborenen", in: *Geschichte des Ungeborenen: Zur Erfahrungs- und Wissenschaftsgeschichte der Schwangerschaft, 17.-20. Jahrhundert*, ed. by eadem, Jürgen Schlumbohm, and Patrice Veit, Göttingen 2002, pp. 11–48.

[10] Cf. Steven Shapin and Simon Schaffer, *Leviathan and the air pump. Hobbes, Boyle, and the experimental life*, Princeton 1985; Steven Shapin, *A social history of truth. Civility and science in seventeenth century England*, Chicago and London 1995.

natural philosophy and followed Galen's model, it was in fact constitutive. Even physicians who put great emphasis on observation and experience, usually insisted unconditionally on the necessity of connecting it with *ratio*. Not least of all, the importance of reason was invoked as a means of criticizing the many irregular healers, the "empirici", who, likewise held up experience as essential in their curative practice, but whom the physicians accused of lacking the required well-trained *ratio*.

Radical doubts about the significance of *ratio*—at least in the conventional sense—spread among the Paracelsians and Helmontians, however. These groups advocated a decidedly empirical approach, and so the alchemical laboratory, the sickbed, and in the case of the Paracelsians, they considered experiential knowledge of "old women" and other uneducated healers the basis for theory and practice. Because the unsoundness of human *ratio* seemed more than proven to their minds by the multitude of medical theories and the resulting vehement disputes, they put direct inspiration from God in its place. The divine light of grace, perhaps revealed to the reader as dream visions, as in the case of Helmont, promised direct access to the one medical truth. They invoked God as the highest authority. Such experiences of divine inspiration, however, were reserved for only the few. Moreover, contemporary critics knew all too well from their contact with the different currents of radical Protestantism that even (purportedly) divine inspiration could lead to contradictory insights.[11]

Personal Authority

The increasing importance of empiricism in early modern medicine and natural philosophy was closely intertwined with a new culture of scientific originality. Apart from learned commentary on authoritative writings, there were now other forms of scientific practice and writing, and not least of all new opportunities for beginning and advancing a career in science. Personal discoveries, original thought, and explanatory models were increasingly valued higher than subtle expositions of traditional texts. The

[11] Walter Pagel, *Joan Baptista van Helmont*, Cambridge 1982; Berthold Heinecke, *Wissenschaft und Mystik bei J. B. van Helmont (1579–1644)*, Bern 1996; Michael Stolberg, "Die Vision als Modus der medizinischen Wissensautorisierung. Johann Baptist von Helmont (1579–1644) und sein 'Aufgang der Arzneikunst'", in: *Morgen-Glantz. Zeitschrift der Knorr-von-Rosenroth-Gesellschaft* 13 (2003), pp. 47–72.

novelty and originality of scientific ideas and discoveries attained a value in itself and slowly replaced the hitherto preeminent requirement for agreement with the recognized classical authorities. In the process, the personal and new increasingly became the decisive basis for a scientific reputation, opening doors to renown, prestigious positions, and, of course, material wealth. Names were made with presentations of the rarest observations possible—monstrosities, fasting miracles, humans giving birth to toads, and so forth. Natural philosophers proudly took advantage of their right to name a discovery such as an anatomical structure that was hitherto unknown or undescribed. For the first time in history, in fact, anatomical structures were connected to the names of contemporary anatomists, as in the case of the fallopian tube, the "tuba falloppiana". At times, there were fierce disputes about who had earned the recognition. This even happened in instances we find difficult to comprehend in retrospect, as in the controversy between Colombo and Falloppio about the "discovery" of the female clitoris. Accusations of plagiarism, which earlier generations might not have even understood in this form, were emphatically denied.

In the course of this development and the concomitant relativization of classical authorities, contemporary physicians were increasingly placed next to them as authorities. In anatomy, a towering figure was Andreas Vesal, but Fabrizio d'Acquapendente, Realdo Colombo, Gabrielle Falloppia, Felix Platter, and Caspar Bauhin also achieved acclaim. In clinical medicine, physicians like Giovanni Battista da Monte, Leonhart Fuchs, Jean Fernel, and Daniel Sennert were increasingly considered to be on par with Galen and Hippocrates, and by the seventeenth century, they were even quoted more often in some medical publications.

With respect to practical medicine, which was understood to be at the heart of medicine even by physicians at the time, this raises the question, however, which factors, which strategies, enabled individual physicians to become recognized as preeminent authorities. From today's perspective, there is no reason to believe that patients generally recovered sooner or more readily in the care of a famous physician like Fernel or Sennert than when treated by a less famous physician or any village barber. On the whole, judged by today's standards, the common treatment with emetics, laxatives, and bloodletting did more harm than good. Neither can the numerous, often very complex mixtures of drugs prescribed by physicians be taken to have had a significant influence on the recovery process.

When contemporaries nevertheless credited famous physicians with superior abilities in the treatment of illness and outstanding healing success, it is clear in retrospect that this attribution came as a result of the authority these physicians had acquired, rather than generating it in the first place.

Very little research has been done on the dynamics that led to the rise of new authorities in early modern medicine. I would like to offer some conjectures. In general, one important prerequisite for a physician's rise to international fame was the publication of medical treatises and textbooks that bore all the marks of medical learnedness. The outer appearance of the medical treatise—its expensively designed title page and its typeface and illustrations—already served to underscore the author's authority and significance. Author portraits or a copper engraving on the title page showed him in sumptuous clothing, assuming the pensive, dignified pose of a scholar or a practicing anatomist or natural philosopher, working on a cadaver or pictured among his chemical laboratory apparatuses. All of this accentuated his claim to outstanding personal experience and skill.[12] Naturally, the author also had to meet the needs of his medical readers with the subject of his writing, with its presentation and argumentation. In this context, it becomes clear why the famous, esteemed physicians of the sixteenth and seventeenth centuries usually connected a measure of innovation with a profound command of traditional medical knowledge. As a leading proponent of medical humanism, Leonhart Fuchs, for example, promised to revitalize medicine on a classical foundation by radically eliminating its Arab elements.[13] Jean Fernel enriched Galenic humoral pathology by adding a new, neoplatonic conception of "specific" diseases and the effects of medicines, and a new theory of "spirits".[14] Daniel Sennert for his part carefully went about integrating Paracelsian, iatrochemical elements into Galenic-Aristotelian medicine.[15]

[12] Caspar Bauhin, *De corporis humani partibus externis*, Basel 1592, dedication; on the functions of author's portraits see Wolfgang U. Eckart, "Zur Titelgestaltung medizinischer Fachbucheditionen des 17. Jahrhunderts", in: *Medizinhistorisches Journal* 14 (1979), pp. 292–311, cit. pp. 293–294.

[13] Leonhart Fuchs, *Paradoxorum medicinae libri III, in quibus sane multa a nemine hactenus prodita Arabum aetatisque nostrae medicorum errata non tantum indicantur, sed et probatissimis autorum scriptis, finissimisque rationibus ac argumentis confutantur*, Basel 1535.

[14] Jean Fernel, *De abditis rerum causis libri duo*, Paris 1548.

[15] Daniel Sennert, *De chymicorum cum aristotelicis et galenicis consensu et dissensu*, Wittenberg 1619.

At the same time, successful physicians drew on a broad spectrum of self-staging and "networking" strategies to acquire and maintain a prominent position in the international scientific community. Correspondence with other physicians played an important role here, letters of recommendation and personal visits served to establish relationships and to make material and immaterial gifts which the person on the receiving end would have to reciprocate. When these authors published, they gave due emphasis to their "friendships" with leading luminaries, for example, by printing their letters, by writing about these famous authorities in the preface, by including dedicatory poems penned by them, by describing observations these men had communicated to them personally, or by referring to joint consultations.

Collective and Individual Strategies of Medical Authorization in the Public Space

Early modern physicians had to defend their theories and practices and their personal credibility not only in the internal scholarly discourse, in their exchange with other physicians. Both individually and collectively, practicing at the sickbed and as representatives of their profession, they had to establish and re-establish, claim and reclaim their authority in everyday encounters with patients and relatives in cities, towns, and in the countryside.[16] This medical authority was threatened first and foremost by the multitude of less-educated competing healers who provided medical care for the majority of the population until well into the nineteenth century. In numerous writings, physicians raked those who arrogated themselves to diagnosing and treating illness over the coals: the "old women", the apostate monks and clerics, the market criers, the itinerant students and gypsies, and most of all the "shameful heretics of

[16] On the social context of early modern physicians' practice see Dorothy and Roy Porter, *Patient's progress. Doctors and doctoring in eighteenth-century England*, Cambridge 1989; Robert Jütte, *Ärzte, Heiler und Patienten. Medizinischer Alltag in der frühen Neuzeit*, Munich and Zürich 1991; Laurence W. B. Brockliss and Colin Jones, *The medical world of early modern France*, Oxford 1997; for a survey on the situation in the preceding centuries, see Michael R. McVaugh, *Medicine before the plague. Practitioners and their patients at the Aragonese court*, Cambridge 1993.

medicine", the Jewish physicians.[17] What was worst of all in the physicians' minds: the public held them "in such authority and belief" that "learned physicians often have to yield to these vagrants".[18] With good reason, physicians composed their vociferous criticism in the vernacular or even explicitly dedicated these writings to municipal or territorial authorities.[19] For the sake of the common good, they wanted them to take aggressive measures against the "deceptive activities", the "murderous doings", indeed the "tyranny"[20] of such healers, who "do not know the causes of natural medicine, have not proper experience, and have not received a good foundation for their art, neither through a learned master nor from ancient books".[21]

The authorities, however, remained tentative in their responses to these wishes. The ordinances that were drawn up against irregular healers stayed on paper in many places. Ananius Horer knew of two reasons "why, in this day, this excellent and highly necessary art, glory, and authority has so dwindled that hardly anyone gives it its due honor and respect anymore". At fault was the "obstinate simplemindedness and childish thoughtlessness of the common rabble and all idiots", who, hoping to be healed, believed in anyone who passed himself off as a "medicus", and the lack of magisterial laws and measures against such "quack doctors".[22] Measures to organize and consolidate physicians' efforts also met with only limited

[17] Ludwig von Hörnigk, *Politia medica Oder Beschreibung dessen was die Medici, so wohl ins gemeins als auch verordnete Hof- Statt- Feldt- Hospital- vnd Pest- Medici [...] So dann endlichen: Die Patienten oder Krancke selbsten zu thun*, Frankfurt am Main 1636, dedication to the Frankfurt magistrate; similarly, Johannes Lange, *Epistolarum medicinalium volumen tripartitum*, Frankfurt 1589, book 1, p. 54.

[18] Georg Pictorius, *Von zernichten Artzten*, Strasbourg 1557, fol. IIv.

[19] Wilhelm Adolph Scribonius, "De inspectione urinarum, contra eos qui ex qualibet urina de quolibet morbo judicare volunt, physiologia cursoria", in: idem, *Idea medicinae secundum logicas leges informandae et describendae. Cui accessit de inspectione urinarum*, Basel 1585, pp. 11–57, dedicatory letter to the consuls and senators of Lemgo; Euricius Cordus, *De urinis, das ist, von rechter Besichtigunge des Harns, und ihrem Mißbrauch*. Ed. by Joh. Dryander, Frankfurt 1543, dedication to the mayor and the town council of Braunschweig; Petrus Talpa, *Empiricus, sive indoctus medicus. Dialogus brevis*, Franeker 1595, dedication to the Frisian senators.

[20] Lange, Epistolarum [see note 17], book 1, p. 54.

[21] Clauser, Betrachtung [see note 3]; similarly, Talpa, Empiricus [see note 19].

[22] Ananius Horer, *Artzney-Teuffel, Oder kurtzer Discurs, Darinn diesem Ertzmörder seine Larve abgezogen*, sine loco 1634, p. 21.

success, as for example the demand Hornung placed on his colleagues to assert themselves with the authorities and win their support in acting against sympathetic healers "so as to ward off such disaster in good time, and so that – as much as possible – patients' health and also the authority and dignity of the physician may be maintained".[23] It was only in a few of the larger cities such as Nuremberg that physicians succeeded in organizing themselves into medical colleges to exert more political pressure on the undesirable competition.

Without the support of the authorities, physicians indeed faced significant difficulties in the relatively unrestricted medical market of the early modern period. As is shown by numerous extant patient letters and other ego-documents, patients and their relatives primarily judged healers by their curative success. Medicine was not comprised "like many other arts, of words, but of the act of making [people] healthy", the Wetzlar printer Spieß reminded the physicians.[24] Yet, the crucial weakness of learned medicine lay precisely here. The usual treatment by physicians, which involved bloodletting, emetics and purgatives, artificial ulcers, and complex mixtures of drugs, was not significantly different from that of the less-educated competition. For the most part both—judged by today's standards—were at best ineffective and at worst harmful. There is no doubt that numerous patients experienced an improvement while undergoing medical treatment—just the naturally favorable course taken by many acute illnesses is enough to account for this—and they were sincerely grateful to their physicians for their help and would sing the praises of their curative success.[25] Yet the same was also true for the curative efforts of the other less-educated and less costly healers.

[23] Johannes Hornung, *De uroscopia fraudulenta discursus. Kurtzer Bericht von dem unvollkommenen und betrüglichen Urtheil des menschlichen Borns oder Harns: wider etliche vermässene Ärtzte, welche allerley Leibsschwachheiten darauß ursprünglichen und gewiß zu erkennen sich fälschlich außthun, dadurch den Patienten nicht allein schädlich und ungewiß gerahten ist, sonder auch der rechte Gebrauch der unaußforschlichen Kunst der Artzney verachtet wirdt*, Herborn 1611, p. 9.

[24] Johannes Schenck von Grafenberg, *New Artzney-Buch*, Frankfurt 1608, printer's preface.

[25] Michael Stolberg, "Die wundersame Heilkraft von Abführmitteln: Erfolg und Scheitern vormoderner Krankheitsbehandlungen aus der Patientensicht", in: *Würzburger medizinhistorische Mitteilungen* (2003), pp. 167–177.

Faced with this situation, physicians pursued a strategy that is still at the core of any professionalization process today[26]: they attempted to dissociate as much as possible an assessment of their knowledge and skill and therefore of their individual and collective authority from the concrete results of their work, that is from their curative success. The population, especially patients from the ruling and wealthy upper classes, needed to be convinced that academic medicine was in its very essence—simply because of its particular, scholarly nature—incomparably superior to all other healing arts. With grandiose language, the physicians staged themselves in numerous writings as the proponents of the one and only true medical art—learned medicine. In so doing, they were simultaneously appealing to general ideals of knowledge and education which they shared with the upper classes from which most of them themselves came. To them, medicine was "much too great, difficult, varied, high and extensive an art and neither should nor could be done and practiced by such uneducated fools".[27] If they admitted at all that their less-educated competitors experienced curative success, they dismissed it as merely due to chance.[28] They did not want to give credit to the reputations that some healers had earned for themselves by having supposedly cured patients. Rather, in Euricius Cordus's words "the common rabble's great praising and proclaiming must hardly be heeded"; after all, many people believed in the healing power of images of the Virgin Mary or called for Saint Nicholas to come to their aid when they found themselves in trouble at sea, only to drown all the same.[29]

Learned physicians furthermore underscored the superior, scholarly character of the medicine they practiced in their day-to-day dealings with patients and their relatives. The outward habitus of the physician—the staff he carried and his noble attire ill-suited to manual labor—already testified to this claim, just as did his refined choice of words and his use of technical Latin or Greek terminology at the bedside. By contrast, physicians left the manual tasks of the trade such as bloodletting mostly

[26] Eliot Freidson, *Profession of medicine*, New York 1970.

[27] Cordus, De urinis [see note 19].

[28] Pieter van Foreest, *The arraignment of vrines*. Transl. by James Hart, London 1623, p. 41.

[29] Cordus, De urinis [see note 19].

to surgeons or barbers. Moreover they relied on other strategies of self-representation that strengthened their authority at the sickbed, as had been expounded since the Middle Ages in works such as "De accessu medici", "De cautelis medicorum" and "Machiavellus medicus".[30] The physician's sophisticated diagnostic abilities could be vividly conveyed, for example, by the ritual of a thorough pulse examination lasting many minutes or an extensive inspection of, and commentary on, the patient's urine.

In their explanations and prescriptions at the sickbed, physicians furthermore laid out two central and characteristic aspects of their medicine which they repeatedly emphasized in their polemics against other healers. For one thing they claimed that, thanks to their extraordinarily thorough knowledge of human nature, they were able to draw precise conclusions about the disease process within the body on the basis of external disease symptoms, and thus were able to explain and treat the illness causally. In extant consultation letters, the presumed causal connections and the location of the disease process within the body are dealt with in detail again and again. And it appears that more or less the same was true for personal consultations at the bedside, as is shown by passages in patient letters and autobiographies. A patient might learn, for example that his podagra was caused by "subtle phlegmatic matter with a degree of mucus and an admixture of bile".[31] Or that his asthma was due to a "persistent obstruction of the lungs and arteries" caused by mucus that flowed from the brain which had already hardened to the point that it assumed the consistency of hail or a hard cheese.[32] Patients and their relatives—and this was true across all classes—expected such explanations. They requested to hear the "most distant and original causes" and they talked

[30] Eckart, Auctoritas [see note 2]; Barbara Elkeles, "Arzt und Patient in der medizinischen Standesliteratur der Frühen Neuzeit", in: *Heilkunde und Krankheitserfahrung in der Frühen Neuzeit*, ed. by Udo Benzenhöfer and Wilhelm Kühlmann, Tübingen 1992, pp. 131–143.

[31] Ioannis Baverius, *Consiliorum de re medica sive morborum curationibus liber*. Ed. by Walther H. Ryff, Strasbourg 1543, foll. 160r-169r, consilium for Dr. Ioann. Bapt. a Ungaria.

[32] Jean Fernel, "Consiliorum liber. Cui accesserunt responsa quaedam clarorum medicorum Parisiensium", in: idem, *Universa medicina*, Geneva 1644, pp. 247–397 (appendix to "De abditis rerum"), cit. pp. 297–300, consilium for a patient from Burgos.

to each other about them.[33] An essential competency for physicians was thus their ability to present a differentiated causal explanation. Presumably, this is also the context in which the remarkable rise of anatomy in early modern medicine must be interpreted to some extent. With the exception of surgery, the anatomists' findings, remained largely irrelevant for the diagnosis and treatment of specific diseases. It was already known that the liver was on the right and the spleen on the left. But particularly in the form of a public anatomical spectacle, anatomical dissections impressively conveyed and underscored the physicians' claim to superior knowledge of the inside of the body, while to this kind of self-staging was largely barred to the competition.

Secondly, physicians boasted that they, unlike their competitors with their supposed "specifics" and universal remedies, thoroughly considered the individual physical constitution of the patient—including age, sex, temperament, and living conditions—and that they adjusted their treatment accordingly. They were persistent in pointing out that only the learned physician was able to do this since it was necessary to take multiple factors into account simultaneously, to weigh them against each other, and to devise a corresponding treatment. Physicians also answered their patients' expectations in this respect—or created them. Educated patients were explicit in their demands for dietetic advice and prescriptions of medicines that were adjusted to their individual constitution. And they hesitated to consult a different healer whom they knew less well because he would be less familiar with their constitution. This markedly individualizing approach helped physicians retain their patients. However, it also made it particularly difficult to arrive at general and binding guidelines for the appropriate treatment of different diseases. As patients lamented time and again, it was their frequent and painful experience that different physicians arrived at very different conclusions when asked about the same case of illness. The authority of learned medicine was considerably jeopardized by this, especially when physicians stood at the sickbed and argued with each other in the presence of patients and relatives.

[33] Cordus, De urinis [see note 19]; for a detailed study of early modern medical lay theories see Michael Stolberg, *Homo patiens. Krankheits- und Körpererfahrung in der Frühen Neuzeit*, Weimar 2003.

Conclusion

Personal self-fashioning and rhetorical-discursive strategies of authorizing knowledge and practices connected seamlessly in early modern learned medicine and in fact overlapped at times. This is true as much for the interactions and discussions among physicians as it is for the claim to, and defense of, medical authority in their dealings with patients and other, competing healers. It is difficult from today's vantage point to assess the success and relative significance of the various strategies. Physicians could attain an eminent position within the scientific community through different life and career paths. And their public self-staging as noble "scholars", while it rendered them vulnerable to criticism and derision when they failed at the bedside, also invited imitation: physicians complained that even simple uroscopists bathed in an aura of esoteric knowledge, cultivating impressive diagnostic and therapeutic rituals, and providing detailed explanations of illnesses based solely on a visual inspection of urine.[34]

Remarkable in any case is the long-term success of early modern physicians' self-representation. Even though, in today's estimation, their diagnoses and methods of treatment were not superior to those of their competitors, they were able to establish themselves among the educated and solvent patients they coveted, becoming their preferred healers. At all times, the authority of any given physician at the sickbed was imperiled by the possibility of therapeutic failure, false prognosis, and medical judgments that differed from those of colleagues. The conviction, however, that "learned" medicine was fundamentally superior to other healing practices emerged largely victorious among the upper classes, helping many physicians attain social status, political influence, and not least of all economic prosperity.

[34] Michael Stolberg, Medizinische Deutungsmacht und die Grenzen ärztlicher Autorität in der Frühen Neuzeit", in: *Macht des Wissens. Entstehung der modernen Wissensgesellschaft 1500–1820*, ed. by Richard van Dülmen and Sina Rauschenbach, Cologne and Weimar 2004, pp. 113–130.

Between Identity-Formation and Self-Staging: Medical Self-Fashioning in the Early Modern Period

The early modern period commonly is considered to be one in which, especially in urban societies, individuals were now more at liberty to live and shape their lives according to their own ideas and desires.[1] The history of learned physicians and of the medical profession is paradigmatic for this development.

In the fifteenth century, educated physicians still constituted a small minority north of the Alps. Even in the more important cities, the municipal "physician" was often a craftsman-surgeon, a "master", not a *physicus* holding a doctorate.[2] This would change fundamentally in the course

First published in German as "Zwischen Identitätsbildung und Selbstinszenierung. Ärztliches Self-Fashioning in der Frühen Neuzeit", in: *Diskurse – Körper – Artefakte. Historische Praxeologie in der Frühneuzeitforschung*, ed. by Dagmar Freist, Bielefeld: transcript-Verlag 2015, pp. 33–55. Translated into American English by Logan Kennedy and Leonhard Unglaub.

[1] Richard van Dülmen, *Die Entdeckung des Individuums 1500–1800*, Frankfurt 1997.

[2] Useful survey in Martin Kintzinger, "Status medicorum. Mediziner in der stadtischen Gesellschaft des 14. bis 16. Jahrhunderts", in: *Städtisches Gesundheits- und Fürsorgewesen vor 1800*, ed. by Peter Johanek, Cologne 2000, pp. 63–92; on the similar situation

M. Stolberg, *Medicine and the Body in Early Modern Europe*, Palgrave Studies in Medieval and Early Modern Medicine,
https://doi.org/10.1007/978-3-031-84556-7_7

of the sixteenth century. The number of academically educated physicians increased dramatically and the number of places where at least one learned physician was active multiplied.[3] The great majority of patients continued to be treated by non-academic healers—by barber surgeons and barbers who were trained as craftsmen, by apothecaries and not least of all, by numerous "wise-women", shepherds, executioners, itinerant merchants, and other irregular healers. However, learned physicians increasingly succeeded in establishing a belief in the fundamental superiority of academic medicine, at least among the educated and wealthy, that is among those who secured physicians' professional, economic, and social advancement when they extended their trust and offered their endorsement. Municipal authorities now filled vacant town physician positions almost exclusively with physicians who held doctorates, and they sometimes went to great lengths to obtain a *doctor medicinae* for this position, which entailed caring for the poor, protecting against plagues and providing medical opinions to the court of justice. Members of the nobility and princes largely preferred to have *doctores* as their court physicians. In their employment contracts, town physicians were given a supervisory role over apothecaries and other medical personnel. Following the Italian model, some cities tolerated or even encouraged the establishment of *collegia medica*, whose physician members decided who was allowed to practice in the city—and thus be in competition with them.[4]

in the Netherlands, see J. van Herwaarden, "Medici in de Nederlandse samenleving in de late Middeleeuwen (veertiende-zestiende eeuw)", in: *Tijdschrift voor Geschiedenis* 96 (1983), pp. 348–378; W. Frijhoff, "Non satis dignitatis.... Over de maatschappelijke status van geneeskundigen tijdens de Republiek", in: *Tijdschrift voor Geschiedenis* 96 (1983), pp. 379–406; M. J. van Lieburg, "Over de stadsmedicus in de Noord-Nederlandse steden van de 16e eeuw", in: *Pieter van Foreest. Een Hollands medicus in de zestiende eew*, ed. by H. L. Houtzager, Amsterdam 1989, pp. 41–72.

[3] A systematic prosopography of early modern physicians remains a major desideratum. I am drawing here on my preliminary analysis of the entries on currently about 4.000 physicians in the bio-bibliographical database which we have been creating, in Würzburg since 2009, under the auspices of the Bayerische Akademie der Wissenschaften, in a long-term project on "Early modern physicians' letters" (www.aerztebriefe.de).

[4] The most important examples north of the Alps are the *collegia medica* in Augsburg and Nürnberg. On the controversies regarding their aspirations and rights see Karl Gröschel, Des Camerarius Entwurf der Nürnberger Medizinalordnung 'Kurtzes und ordentliches Bedencken' 1571. Diss. med., Munich 1977; ample documentation of the activities of the local *collegium medicum* has survived in the Stadtarchiv Augsburg.

With their professional activities and the central position they enjoyed—in particular with respect to providing medical care for the educated upper classes—and with the accompanying opportunity to earn a good living, many physicians could walk the road to status and esteem. Certain physicians, Joachim Vadian in St. Gallen for example, even held important political offices in their municipality.[5] Only few physicians, like Joachim Camerarius II for example, had fathers who were influential and extremely well connected, or perhaps even came from medical dynasties such as that of the Occos in Augsburg or the Bauhins in Basel. Most physicians practiced medicine far away from the town where they had grown up. In these circumstances, then, higher education in Latin schools, medical studies, and a doctorate became the all-important keys that opened the door to social advancement, even for the sons of cobblers, goldsmiths, or brewmasters. Thus, it can be said that physicians of this time are representative of the growing number of educated people whose wealth and higher social standing were due almost exclusively to their university education and their own labor.

The functions physicians could fulfill and the fields of activity in which they could engage as a result of their academic education were varied. Not seldom did physicians hold—in sequence but sometimes even simultaneously—various positions and offices in different contexts in the course of their careers. Most widespread was independent practice, generally in larger towns, which promised wealthy patients in sufficient numbers. Many physicians accepted the position of town physician, at least temporarily, or they were responsible, sometimes at the same time, for health care at municipal hospitals or other hospitals such as the "Franzosenhäuser" that specialized in the treatment of patients with the "French disease".[6] Some physicians were appointed university professors, sometimes directly after they had earned their degree, as in the case of the

[5] Bernhard Milt, *Vadian als Arzt*, St Gallen 1959.

[6] On the figure of the town physician, see the contributions to Andrew W. Russel (ed.), *The town and state physician in Europe from the Middle Ages to the Enlightenment*, Wolfenbüttel 1981; Ruth Schilling, Sabine Schlegelmilch, and Susan Splinter, "Stadtarzt oder Arzt in der Stadt? Drei Ärzte der Frühen Neuzeit und ihr Verständnis des städtischen Amtes", in: *Medizinhistorisches Journal* 46 (2011), pp. 99–133; for well-researched case studies on physicians in Nürnberg, see Klaus G. König, *Der Nürnberger Stadtarzt Dr. Georg Palma (1543–1591)*, Stuttgart 1961, and Peter Assion and Joachim Telle, "Der Nürnberger Stadtarzt Johannes Magenbuch. Zu Leben und Werk eines Mediziners der Reformationszeit", in: *Sudhoffs Archiv* 56 (1972), pp. 353–421.

shoemaker's son Daniel Sennert in Breslau.[7] A position that was particularly lucrative and sought-after, though it was accompanied by a distinct relationship of dependency, was the position of *Leibarzt* or court physician for a high-ranking person, a king or a prince or indeed the emperor. Even a prominent scholar like Andreas Vesal, who became famous all across Europe upon the publication of his *De humani corporis fabrica libri septem*, decided not to continue with his research when the opportunity arose to become imperial court physician. He resigned from his professorship, took the new position, and as good as vanished from the world of research and publication.[8]

Whether they limited themselves to private practice, worked as court physicians, or fulfilled public duties as town physicians, most physicians had one thing in common as far as we can tell from the extant sources: for their whole life, they maintained and lived according to a learned habitus.[9] In particular, physicians played a very important role in the lively epistolary communication that took place within the contemporary Republic of Letters.[10] There are tens of thousands of extant letters penned by physicians in the sixteenth and seventeenth centuries.[11] Some physicians, such as Crato von Krafftheim (1519–1585) and Caspar Peucer (1525–1602) carried on such bountiful correspondence that letter writing would have taken up a significant portion of their working day.[12] Quite

[7] "Vita Danielis Sennerti", in: Daniel Sennert, *Opera omnia*, Lyon 1656 (no page numbers).

[8] On Vesal's biography see Charles D. O'Malley, *Andreas Vesalius of Brussels 1514–1564*, Berkeley and Los Angeles 1965.

[9] It is important to keep in mind, however, that sources on well-connected learned physicians are much more likely to have survived than those on the odd town physician who concentrated largely on taking care of his patients.

[10] For an exemplary case study, see Tilmann Walter, "Ärztliche Selbstdarstellung im Zeitalter der Fugger und Welser. Epistolarische Strategien und Repräsentationspraktiken bei Felix Platter (1536–1614)", in: *Personen und Milieu. Individualbewusstsein? Persönliches Profil und soziales Umfeld*, ed. by Angelika Westermann and Stefanie von Welser, Husum 2013, pp. 285–314.

[11] The data base of the Würzburg project "Early modern physicians' letters" (see note 3) currently has entries on about 20.000 letters which were written by or addressed to physicians in the sixteenth- and seventeenth-century German speaking areas; about 15.000 more have already been identified [2024: the number has increased to altogether about 65.000 letters].

[12] Johann Franz Albert Gillet, *Crato von Crafftheim und seine Freunde. Ein Beitrag zur Kirchengeschichte*, 2 parts, Frankfurt am Main 1860; Hans-Peter Hasse and Günther

often, the letters concerned medical questions in a narrower sense, and sometimes they were even about specific cases, about shared patients. In many letters, however, the physicians took up topics that lay far beyond the bounds of their discipline and in a broader sense were commensurate with humanist erudition—everything from genealogy and local history to linguistic research to numismatics.[13] Many a physician also wrote poems. Often, the poems are laudatory, dedicatory, or otherwise catered to a particular occasion, as we often find them as paratexts in medical works. But there were also very successful and famous poets among medical practitioners, for example Johannes Posthius (1537–1597) and Petrus Lotichius (1528–1560). Even the short entries in contemporary friendship books or *alba amicorum*, show how medical students and physicians were underscoring their knowledge of the cultural heritage of antiquity, of the philosophers, poets, and historians; sometimes they even immortalized themselves by writing quotes in Greek.[14] Some physicians went so far as to collect such passages and quotes for later use in letters, poems, forewords, and so forth in the form of handwritten *loci communes*, the typical scholarly practice of notetaking at the time.[15] Approximately 1100 pages with such *loci communes* on non-medical topics penned by the Bohemian physician Georg Handsch (1529–1578?), for example, have survived.[16] A strikingly large number of physicians distinguished themselves with contributions to historical research.[17] Some

Wartbergs (eds), *Caspar Peucer (1525–1602): Wissenschaft, Glaube und Politik im konfessionellen*, Leipzig 2004.

13 Universitätsbibliothek Erlangen, Trew Collection, letter from Adolph Occo III to Joachim Camerarius II, Augsburg, February 12, 1575, on his numismatic research and his intentions to publish the results.

14 See, for examle, National Library of Medicine, Bethesda, Ms E77, *Album amicorum* of Conrad Gessner.

15 Staatsbibliothek Berlin, Ms. Lat. Qu. 41, humanistic *loci communes* compiled by Salomon Alberti; on the genre, see Michael Stolberg, "Medizinische *Loci communes*. Formen und Funktionen einer ärztlichen Aufzeichnungspraxis im 16. und 17. Jahrhundert", in: *NTM - Zeitschrift für Geschichte der Wissenschaften, Technik und Medizin* 21 (2013), pp. 37–60.

16 Österreichische Nationalbibliothek, Vienna (henceforth: ÖNV), Cod. 9550, *Promptuarium sive loci communes latinitatis* by Georg Handsch (around 1550).

17 Gianna Pomata and Nancy G. Siraisis (eds), *Historia. Empiricism and erudition in early modern Europe*, Cambridge, MA, 2005; Nancy G. Siraisi, *History medicine and the traditions of Renaissance learning*, Ann Arbor 2007.

physicians worked intensively on natural-philosophical subjects and or even became known primarily as natural philosophers. They conducted mineralogical or zoological studies, like Georg Agricola (1494–1555) and Conrad Gessner (1516–1565), or they described, as did the aforementioned Georg Handsch, modern methods of fish farming.[18] Others made a name for themselves as astrologers and astronomers. Nicolaus Copernicus (1473–1543) and Girolamo Cardano (1501–1576) are two famous examples. The young Johannes Magirus (1615–1697) concerned himself with fortress construction.[19] Heinrich Stromer alias Dr. Auerbach (1476–1542) was the author of a book about mathematics that went through several editions.[20]

Learned habitus, the claim to partake in the contemporary *res publica litteraria*, also found expression iconographically. When educated physicians had likenesses made of themselves, for example author portraits, they commonly did not show themselves at the sickbed, which is to say in the place where they practiced their medical profession more than anywhere else, but rather in their study at home, surrounded by books, writing implements, and other attributes of erudition. This (self)representation of the physician, who had commissioned the portrait and thus governed its rendering, was clearly distinct from depictions of physicians in contemporary genre painting, which almost always showed them practicing uroscopy and/or pulse diagnosis at the bedside.[21]

Physician Self-Fashioning

At closer analysis, the learned self-understanding and learned habitus of physicians of the period is surprising. After all, the overwhelming majority of physicians earned their living from medical practice, from treating

[18] See the modern edition of the relevant passages in Handsch's manuscript *Historia animalium* in Georg Handsch, *Die Elbefischerei in Böhmen und Meißen*. Ed. by Ottokar Schubert, Prague 1933.

[19] Universitätsbibliothek Marburg, Ms. 392, excerpts on mathematics and other topics by the physician Johannes Magirus (mid-seventeenth century).

[20] Heinrich Stromer, *Algorithmus linealis numerationem, additionem, subtractionem, duplationem, mediationem, multiplicationem, divisionem et progressionem una cum regula de tri perstringens*, Leipzig 1504.

[21] Kitti Jurina, *Vom Quacksalber zum Doctor medicinae. Die Heilkunde in der deutschen Graphik des 16. Jahrhunderts*, Cologne 1985, esp. pp. 34–37; Susanne Fürst, *Das Arztporträt in der Frühen Neuzeit*. Diss. med., Regensburg 2009.

patients. Even those who distinguished themselves as fortress builders, astrologers, or poets could only in extremely rare cases feed a family or send their sons to university from these sources of income. In the great majority of cases, the economic advancement of town physicians and professors, whose salaries as such were modest, also depended crucially upon earnings from medical practice. Annual incomes of 60 or even 120 gulden, as we find in the employment contracts of town physicians,[22] were nothing to write home about—especially since, as some town physicians complained, the money was not reliably paid to them. Unlike at the large Italian universities, a professorship north of the Alps did not promise riches either. And even the court physicians of great princes were not always remunerated with a princely sum.[23]

Why, then, did so many physicians nevertheless cultivate a learned habitus in these circumstances? Why, according to their own testimony, did they spend many late hours writing letters and engaging in scholarly study? Was all of this merely self-staging? Did they hope to gain recognition in this way and ride on its coattails to a more lucrative position, perhaps becoming the court physician to a great prince? Or was it rather that this learned habitus expressed their "true" identity? Was it the result of years of shaping, forming, educating themselves? Was there essentially no other way for them to cast themselves than as learned, humanist subjects?

There is evidence for either assumption, but ultimately—and this will become more apparent shortly—the question is based on a false dichotomy. Self-formation and self-staging were inextricably linked. They were the central and in some respects complementary aspects of medical "self-fashioning", a concept the American literary scholar Stephen Greenblatt introduced years ago.[24] Greenblatt's starting point was the finding

[22] See, for example, Institut für Stadtgeschichte, Frankfurt, Dienstbriefe 978, employment contract for Bechtold Bach as town physician with an annual salary of 60 Gulden und 15 Batzen February 28, 1589; Stadtarchiv Memmingen, draft of an employment contract for David Stegmann as town physician, 1621, with an annual salary of 60 Gulden and an additional 15 Gulden every three months and some beech wood.

[23] Staatsarchiv Würzburg, Standbuch 794, copy of an employment contract for Johannes Posthius as personal physician to the prince bishop, February 22, 1569, with an annual salary of 100 Taler (about 115 Gulden); for a detailed summary by Ulrich Schlegelmilch see www.aerztebriefe.de/id/00019762.

[24] Stephen Greenblatt, *Renaissance self-fashioning. From More to Shakespeare*, Chicago and London 1980.

that the term "fashioning" was widely used in the sixteenth century "as a way of designating the forming of a self". Here, a new faith in the possibilities of fashioning one's own life was finding expression or rather, put more generally, "an increased self-consciousness about the fashioning of human identity as a manipulable, artful process".[25] With the concept of "self-fashioning", Greenblatt attempted to grasp the ultimately inextricable entanglement of subjectivity and society, which to this day preoccupies historians,[26] between the individuals who experience himself or herself as the fashioner of his or her own identity, and the respective societal and cultural context, which inevitably and crucially determines the shaping of this identity or subjectivity. Greenblatt summarized the findings of his study with the words, "Fashioning oneself and being fashioned by cultural institutions – family, religion, state – were inseparably intertwined".[27]

In gauging how much autonomy certain sixteenth-century English literary figures enjoyed—and not only minor ones but even leading luminaries—Greenblatt, referring explicitly here to the cultural-anthropological work of Clifford Geertz, arrived at the very skeptical conclusion, which was typical of his time, that, "If there remained traces of free choice, the choice was among possibilities whose range was strictly delineated by the social and ideological system in force".[28] In newer historical research, what emerges by contrast, as the editors of a volume on "self-formings" recently put it, is "the tendency [...] to develop a new, revised understanding of the subject, which takes into account (once again) central dimensions of subjectivity such as agency, capacity for reflection, self-reference, and identity, *without*, however, losing sight of the knowledge of historicity and the societal boundedness

[25] Ibid., p. 2.

[26] For a historiographic survey see Nicolaus Buschmann, "Persönlichkeit und geschichtliche Welt. Zur praxeologischen Konzeptualisierung des Subjekts in der Geschichtswissenschaft", in: *Selbst-Bildungen. Soziale und kulturelle Praktiken der Subjektivierung*, ed. by Thomas Alkenmeyer, Gunilla Budde, and Dagmar Freist, Bielefeld 2013, pp. 125–149.

[27] Greenblatt, Renaissance self-fashioning [see note 24], p. 256.

[28] Ibid.

of the subject".[29] Greenblatt's concept of "self-fashioning" is nevertheless of interest to historical research into the identity of the members of a professional group such as physicians not only for its contribution to understanding more precisely the connection between identity formation and culture. It also points out a second interrelation that is central to comprehending professional identities in particular: that between self-formation and self-staging. "Self-fashioning" as Greenblatt conceptualized it refers not only to the process of subjectivizing but also to a representation of the self to the outside world, to the formation of a public "I", to the "representation of one's nature or intention in speech or actions"—indeed, to "hypocrisy or deception".[30] If we look at this from a praxeological perspective, we can say that self-formation is communicated decisively through a person's practices.[31] This implies that those practices—linguistic and non-linguistic alike—of which members of a professional group avail themselves in order to effectively put themselves "on stage" also have a decisive influence on the way they see themselves, on their identity, and are indeed entangled with it.

With all of this in mind, it is my aim in the following to look primarily at the Germanic lands between 1500 and 1650, to give a brief overview of the essential aspects of professional self-fashioning for early modern physicians as it unfolded between self-formation and self-staging, and to explore the subtle but in the long term very powerful changes that become apparent here.

Learned Habitus

There is much to suggest that the learned habitus consistently cultivated by physicians both grew out of and reflected years of socialization and the concomitant carrying out of learned practices. An essential foundation was laid through intensive schooling in the Latin language, from childhood on. The Bohemian physician Georg Handsch would thank his teacher in Leipa, where he grew up, with the words, "Formasti mihi ad

[29] Editors' "Einleitung", in: *Selbst-Bildungen. Soziale und kulturelle Praktiken der Subjektivierung*, ed. by Thomas Alkemeyer, Gunilla Budde, and Dagmar Freist, Bielefeld 2013, pp. 9–30, cit. p. 9.

[30] Greenblatt, Renaissance self-fashioning [see note 24], p. 3.

[31] Cf. Thomas Alkemeyer, "Subjektivierung in sozialen Praktiken. Umrisse einer praxeologischen Analytik", in: idem et alii, *Selbst-Bildungen* [see note 29], pp. 33–68.

altiora latos gressus".[32] At leading Latin schools, for example in the Silesian town of Goldberg, students were even said to have conversed with one another in Latin in their spare time.[33] There were fathers, such as the physician Wolfgang Reichart, who corresponded with their young sons in Latin.[34] This language training simultaneously gave students a thorough knowledge of the works of authoritative ancient Roman and to some extent Greek authors. In the arts faculties, that is, during an education in the liberal arts, this knowledge was deepened and competencies were improved. Linguistic, logical, and rhetorical skills continued to be taught at these institutions, and a conversance in classical tradition was further enriched. Universities also offered teaching in music, natural philosophy, astronomy, and mathematics, though the weight given to each differed from place to place. Thus, we see how future physicians, earning their baccalaureate or master's degree at a good twenty years of age, had already spent more than a decade studying the scholarly Latin heritage. They were "humanists"—educated in the *studia humanitatis*.

Studying medicine was at this point only one option of many, and one that because of its high cost was certainly not open to everyone. This is made exceedingly clear by a rather intriguing list of professions that appears on the first page of one of the extensive handwritten notebooks left behind by the Bohemian physician Georg Handsch (1529–1578). Handsch, at the time a young man some nineteen or twenty years old, enumerated different occupations or fields of work: "poeta", "orator", "arithmeticus", "musicus", "grammaticus de variis rebus (sic)", "medicus", "organista", and "nigromanticus". Later, it seems, Handsch added "dialecticus", "lector", and "praestigiator". The meaning of this list is illuminated in further entries on the following two pages. He hoped his *magister* would recommend him as *arithmeticus* for the metal works of Herr Gendorf. He also asked that the "Herr"—presumably not a human "Lord" but the Lord God—"may see to it that I become a lector at the university in Prague". This is written in German between two Latin

[32] ÖNV, Cod. 9821, foll. 24r-26v, autobiographic poem "ad praeceptorem" ("Agrestem quoniam imbuisti mentem praeceptis, rudimenta prima monstrans, formasti mihi ad altiora latos gressus"; later Handsch changed "rudimenta" to "elementa").

[33] Gustav Bauch, *Valentin Trozendorf und die Goldberger Schule*, Berlin 1921.

[34] Walther Ludwig, *Vater und Sohn im 16. Jahrhundert Der Briefwechsel des Wolfgang Reichart genannt Rychardus mit seinem Sohn Zeno (1520–1543)*, Hildesheim 1999; the son, Zeno, also became a physician.

entries. A further entry says "Become a town scribe".[35] There is little room for doubt: young Handsch was sketching out possible professional options for himself. The spectrum could have hardly been wider, but, with the exception of "magician" (praestigiator)—Handsch knew some magic tricks—and perhaps of "necromancer" all of his options were ultimately founded on his liberal arts education. In the end, it appears that he was only able to study medicine because he earned the goodwill of the archducal court physician Andrea Gallo, who paid for his studies because he wanted Handsch to go to Padua with his son.[36]

With the exception of some sons of physicians, most future physicians were in no way predestined to this profession. Even for those who were willing and able to complete studies in one of the three higher disciplines following the study of the *artes*, there was the choice between medicine, theology, and jurisprudence. In contemporary correspondence, we thus find discussions about whether one's son or a different promising young man should pursue a medical career, or whether he should perhaps study jurisprudence or theology instead.[37] And there were some who decided to study medicine later only and first worked as a *Hofmeister*, that is, as a tutor to the sons of princes or noblemen, or served, like Georg Agricola in Zwickau, as a school master. The Protestant Reformation in particular opened new perspectives here.

The study of medicine, then, largely built on the scholarly knowledge and skills that had been previously acquired, and thus further contributed to solidifying the acquired scholarly habitus. The first goal was to gain a comprehensive knowledge of the authoritative ancient and medieval texts and of their modern commentators and the ability to use logical arguments to contend with the contradictions and open questions that

[35] ÖNV, Cod. 9666, fol. 1r; Handsch indicated September 23, 1547, as the starting point of his note-taking in this manuscript. He probably added the list at a later point in time but must have done so before he left Prague in 1551 to study medicine in Padua.

[36] So far, Handsch's (few) biographers have assumed that Handsch had secured financial support from a nobleman. In his list of career options cited above Handsch wrote, however: "Doctor Gallus vult me mittere in Italiam cum filio suis sumptibus". After his return to Prague, Handsch lived and worked with Gallus. Handsch's case is somewhat untypical insofar as he apparently only studied at the Charles-Universität in Prague but did not obtain a degree. When he later acquired a medical doctorate in Ferrara, he pretended that he had a master's degree.

[37] Ludwig, Vater und Sohn [see note 34], pp. 224–227, letters from Wolfgang to Zeno Reichart, Feburary 22 and 24, 1524.

these texts raised. It was common practice for lectures to be dedicated to in-depth expository commentary of such texts. In public *collegia* and disputations, students were schooled in the art of scholarly debate.[38]

Learned Self-Staging

We have seen, then, how future physicians were introduced to the contemporary scholarly culture and its practices throughout their entire youth and how they acquired a learned habitus—often with no clear idea what they were going to do for a living in the future. We may thus assume that the learned habitus of university-trained physicians was deeply ingrained and was an essential part of their identity. However, it was also clearly at variance with their professional activity. In their daily work of diagnosing and treating illnesses, they moved in a world that was profoundly different from that of their academic erudition. It was not scholarly writing, but instead the creaturely body that was the center and focus, as well as the attempt—often an all too futile one—to relieve patients of their suffering, to save their lives. Day in, day out, physicians were confronted with lamentations and cries of pain.

And worse yet, in a time in which having contact with "impure" matter was greatly stigmatized, they constantly had to deal with sickening discharge, with sores and ulcers, skin rashes and abscesses, and with their patients' foul-smelling excretions. At the time, most diseases were attributed to the foul, spoiled, burnt, or otherwise harmful morbid matter that accumulated in the body. Because, according to the general conviction, the body tried to rid itself of this morbid matter by way of excretions, taking a close look at them provided valuable clues about the nature of the morbid matter in question. By far the most important diagnostic procedure was uroscopy. In the case of stool inspection, the physician had to literally stir and lift the stool with a small stick to be able to recognize,

[38] C. D. O'Malley, "Medical education during the Renaissance", in: *History of medical education*, ed. by idem, Berkeley 1970, pp. 89–102; Jerome J. Bylebyl, "The school of Padua: humanistic medicine in the sixteenth century", in: *Health, medicine and mortality in the sixteenth century*, ed. by Charles Webster, Cambridge 1979, pp. 335–370; Nancy Siraisi, "The faculty of medicine", in: *A history of the university in Europe*. Vol. I: *Universities in the Middle Ages*, ed. by Hilde de Ridder-Symoens, Cambridge 1992, pp. 360–387; eadem (ed.), *Medicine and the Italian Universities 1250–1600*, Leiden 2001; Vivian Nutton, "Medicine at the German universities, 1348–1500. A preliminary sketch", in: *Würzburger medizinhistorische Mitteilungen* (1997), pp. 173–187.

for example, admixtures of morbid mucus. And medical treatment for its part generally aimed at supporting the body in its excretory efforts: diuretic medicines, laxatives, and/or emetics were the basis of almost all treatments—and here too, it was necessary when determining the effectiveness of the cure to examine the excrement carefully, which perhaps showed an unappetizing, foul-smelling substance that was then assumed to be the morbid matter.

Such activity was difficult to reconcile with the dignity of a learned physician. Agrippa von Nettesheim put his finger on the wound: the medical art, in his devastating assessment, was "filthy". It was only "because of a shameful profit" that physicians "circled around sick people's piss-pots and outhouses". They were "for the most part contagious and reeking of patients' urine and feces", "filthier than even the midwives, as they have to look at nasty and filthy things with their own eyes, and hear and smell the belching and farting of the patients".[39] Young Zeno Reichart was "not born for stool and urine" a befriended apothecary argued along these lines, and recommended that he study law rather than medicine.[40]

It was not only the brutal reality of the diseased or wounded body and the inevitable contact with excrements that stood in open contradiction to the learned self-understanding of physicians. The people, too, with whom physicians dealt on a daily basis—their patients and the relatives of these patients—often belonged to a different world altogether. For a long time, historians believed that, until sometime in the eighteenth century, learned physicians almost exclusively treated noble, wealthy, and hence educated patients.[41] It has been known for some time, however—thanks to practice diaries, medical notebooks, and published case histories—that many tradesmen, farmers, merchants, and servants also sought the advice of learned physicians and even constituted the large majority of

[39] Agrippa von Nettesheim, *Die Eitelkeit und Unsicherheit der Wissenschaft und die Verteidigungsschrift*, 2 vols. Ed. by Fritz Mauthner, Munich 1913, vol. 2, p. 79.

[40] Ludwig, Vater und Sohn [see note 34], letter from Wolfgang to Zeno Reichart, February 24, 1524 ("non ad stercora et lotia esse natum").

[41] Nicholas D. Jewson, "Medical knowledge and the patronage system in 18th century England", in: *Sociology* 8 (1974), pp. 369–385.

their patients. In other words, in their daily practice physicans often were dealing with people who could not properly read or write.[42]

It was this "audience" that determined a physician's reputation and, as a result, his economic advancement. After all, in the largely unregulated early modern health market, it was possible for patients and their relatives to consult or switch to a different physician or less-educated healer at any time if they were unsatisfied with the medical treatment they had received. Which they did. The treasured image we have of the paternal family physician who cared for his patients from the cradle to the grave in centuries past is a fiction. Such fidelity was the exception. In medical practice diaries, most patients appear only sporadically, having had a single consultation or at most having been seen for a few days or weeks. Frequently and quite often in a short time, patients switched from one physician or healer to another.[43]

Working in these circumstances, the physician had to earn the trust of patients and relatives again and again. Ideally, he accomplished this by presenting convincing proof of his superior art: since he demanded a higher fee, patients could expect that he had a better success rate compared to the less-educated competition. Yet it was precisely in this respect that physicians came up against their limitations. They themselves were convinced of the fundamental superiority of the medicine they practiced, and when they were ill themselves, they accordingly turned to other learned physicians.[44] They believed that, buttressed by knowledge that had been accumulated over centuries through medical scholarship, and through the application of their academically trained reason, they could arrive at a much more efficacious therapy than their less-educated competitors and a therapy that was more closely adapted to the specific patient and his illness.[45] In daily practice, however, in their treatment of

[42] Universitätsbibliothek Marburg, Ms. 96, medical diary of Johannes Magirus (around 1650); Stadtbibliothek Ulm, H Franc. 8a and 8b, collection of case histories by Johannes Frank; ÖNV, Cod. 11183, Cod. 11205, Cod. 11206, Cod. 11207, Cod. 11238 and Cod. 11240, notes on medical practice by Georg Handsch.

[43] Kungliga Biblioteket, Stockholm, X 101, *Medicinisches Receptur-Diarium* by Dr Petrus Kirsten 1612–1616; Magirus, Medical diary [see note 42].

[44] Georg Handsch, for example, described various diseases that he himself or medical colleagues of his suffered from and which prompted them to seek the advice of other physicians.

[45] Magirus, Medical diary [see note 42].

patients, they often could not provide proof of their superiority. In no way did the medical results of their practice appear obviously superior to those of other healers. When we look back today and apply for a moment the standards of modern medicine, this is not surprising. The bloodletting, the strong laxatives and emetics that were used for most illnesses would be considered harmful for the most part in today's assessment, and remedies that we would retrospectively consider to have a specific effect for specific ailments hardly existed. It is true that many patients recovered under medical treatment: as we know today, acute, febrile diseases—and these were widespread at the time—often have a positive outcome, no matter how they are treated, and that which we know today as the placebo effect may have favored such outcomes. But we can assume the same was true when we look back on the "healing successes" of other, less-educated healers. They, too, relied heavily on "evacuative" remedies and procedures, and they too had many satisfied patients.

In this situation, the claim to erudition, staged in such a way that it was convincing to the public, constituted an essential mark of distinction. Wearing this badge, learned physicians could hope that they would stand out as members of a professional group which, regardless of the level of success experienced with specific patients, had greater expertise in medicine over and above the many competitors. The general academic title of *doctor*, which ultimately, as "Herr Doktor", became synonymous with the university-educated physician, the *doctor medicinae*, was the crucial attribute that signaled such superiority to the outside world—even if the doctoral examination, apart from its high cost, generally did not represent a particularly difficult obstacle. With good reason, physicians would regularly add this title to their signatures on letters, for example. Conversely, the lack of a doctoral degree served the new colleges of physicians as a decisive criterium for rejecting out-of-town physicians who were seeking permission to practice.[46]

It was not only with their title, but also in their interactions with patients and relatives that physicians sought to stage themselves as scholars. At least this emerges from letters written by physicians to wealthy patients who consulted them from a distance, a practice that was prevalent among the upper classes during this period. The physicians would use technical Greek and Latin terminology, which they could assume

[46] Stadtarchiv Augsburg, archives of the *collegium medicum*.

patients did not understand. They explained patients' diagnoses to them with complex theories and sometimes sprinkled them with quotes from Hippocrates, Galen, or Avicenna, sometimes in the original language, although they presumably were well aware that patients would not be able to deciphter them, let alone verify their accuracy.[47]

Some physicians also excelled at presenting themselves as experts to a wide public. They successfully staged public anatomy demonstrations, like young Felix Platter in Basel, which impressively presented to a large audience their thorough knowledge of the structures in the mysterious interior of the body, which is to say, in the place where most diseases were thought to originate.[48] Others composed advice booklets in the vernacular about proper behavior in times of the plague, about dealing with the French disease, or about a healthy lifestyle in general. Alongside writings on pious edification, medical advice literature was likely the most successful genre in the contemporary book market. And then there were the physician-authors of the widespread astrological-medical calendars and almanacs, who not only catered to a lucrative market, but also presented themselves to a large readership as the bearers of superior knowledge—knowledge that was indispensable to the good of both the individual and the population at large. After all, in order to know which days were best for bloodletting or the administration of medicines, it was necessary to first calculate the planetary positions for each place or region in question.[49] By publishing medical works that promised to be of some use to the common good, one could perhaps also score points in an application for the position of town physician. Marcus Banzer, for example, applied to the Augsburg city council, underscoring that he—unlike many

[47] To quote just a few examples from a wide array of writings: Michael von Glaubitz, *Zwo Haußtaffeln vnd Underricht fur die Reichen vnd Armen zur Sommer vnd Winterzeit wider die fürstehende schreckliche vnd wegfressende Pestilentz*, Mainz 1584; Andreas Starck, *Krancken Spiegel*, Mülhausen 1598; Johannes Wittich, *Praeservator sanitatis*, Leipzig 1590; Johan van Beverwijk, *Schatz der Gesundheit*, Amsterdam 1671.

[48] Michael Stolberg, "Eine anatomische Inszenierung. Felix Platter (1536–1614) und das Skelett der Frau", in: *Spuren der Avantgarde: Theatrum anatomicum. Frühe Neuzeit und Moderne im Kulturvergleich*, ed. by Helmar Schramm, Ludger Schwarte, and Jan Lazardzig, Berlin and New York 2011, pp. 147–167.

[49] See, for example, Philipp Melhofer, *Lasstafel oder Almannach [...] Auff das MDXLIII Jare*, Augsburg 1543; Johannes Magirus, *Alter und newer Schreib-Calender auff das Jahr [...] M.DC.XVI. Auff den Berlinischen und der Marck Brandenburg [...] Horizont fleissig und genau gerichtet*, Altenburg 1646.

others—had not only studied in Montpellier, Padua, and other places, but had also published a book of dispensing instructions, the "Fabrica Receptarum".[50]

The claim to erudition found expression most of all in physicians' scholarly works.[51] By publishing treatises and textbooks, they not only secured their place in the *res publica medica*, but also made a name for themselves among their closer colleagues, who might recommend patients to them or approach them for joint epistolary *consilia*. As a rule, authors flaunted all possible signs of erudition, with countless Latin and sometimes Greek or even Hebrew quotes, with endless references to the ancient medical authorities, and quite often with scattered indications of their wider humanist learning, when they cited Latin poets for example. In sixteenth-century medical humanism with its efforts to arrive at a highly precise, philosophically informed reconstruction of the authoritative ancient texts, these elements of erudition came forcefully to the fore.[52] And even when writing collections of case histories, of *curationes* and *observationes*—a new genre that gained popularity in the second half of the sixteenth century—authors tended to add learned *scholia*, a practice we find especially with the two most important, influential proponents of this genre, Amatus Lusitanus and Pieter van Foreest.[53]

[50] Stadtarchiv Augsburg, Collegium medicum, Karton VIII, 54, letter from Banzer, March 7, 1624; Banzer was referring to his *Fabrica receptarum. Id est: methodus brevis, perspicua ac facilis in qua quae sint remediorum compositorum formae, quae earundem differentiae, quae componendi & praescribendi ratio [...] planissime edocetu*r, Augsburg 1622.

[51] Cf. Michael Stolberg, "Frühneuzeitliche Heilkunst und ärztliche Autorität", in: *Macht des Wissens. Die Entstehung der Wissensgesellschaft*, ed. by Richard van Dülmen and Sina Rauschenbach, Cologne, Weimar, and Vienna 2004, pp. 111–130.

[52] On medical humanism, see in particular Vivian. Nutton, "The rise of medical humanism: Ferrara, 1464–1555", in: *Renaissance studies* 11 (1997), pp. 2–19; Jerome Bylebyl, "Medicine, philosophy and humanism in Renaissance Italy", in: *Science and the arts in the Renaissance*, ed. by John W. Shirley and F. David Hoeniger, Washington, D.C., 1985, pp. 27–49; Wilhelm Kühlmann and Joachim Telle, "Humanismus und Medizin im 16. Jahrhundert", in: *Semper apertus. Sechshundert Jahre Ruprecht-Karls-Universität Heidelberg 1386–1986*, vol. I: Mittelalter und Frühe Neuzeit 1386–1803, ed. by Wilhelm Doerr, Berlin 1985, pp. 255–289.

[53] Amatus Lusitanus, *Curationum medicinalium centuria prima, multiplici variaque rerum cognitione referta*, Florence 1551; Pieter van Foreest, *Observationum et curationum medicinalium libri tres [...] in quibus eorum caussae,signa, prognoses, curatio graphice depinguntur. Etiam adiectis scholiis*, Leiden 1590; over the following years, Foreest published more than 30 further books of medical and surgical case histories.

To all appearances, physicians were successful in staging themselves and their learnedness effectively to the public. We see how, for example, in letters sent to physicians by patients and their relatives, physicians are frequently addressed as "highly learned" ("hochgelehrt").[54] There were, however, limitations to staging oneself successfully as a superior expert. Particularly considering the high fees that physicians demanded, an almost unsurmountable obstacle to becoming cleanly recognized for medical superiority was their frequent (and in retrospect unavoidable) therapeutic failure at the sickbed. What was worse, a physician's self-presentation as erudite could even prove counterproductive in these circumstances and make him the target of biting derision. Medical satire has a centuries-old tradition—the physician figures in Molière's plays are only the most well-known example.[55] This satirizing was fueled above all by the cleft, obvious to all also in the contemporary assessment, that opened between the physicians' claim to erudition—their grandiloquent terminology, their proverbial discussions at the sickbed—and their common failure when making diagnoses and prognoses and treating illnesses.

It was one thing to be ridiculed in literature, but at a political level learned physicians came up against limitations, too, when they sought to obtain the privileges they claimed were their due. Especially in those places where they were able to establish a college of physicians, they were able to secure certain supervisory rights. However, their vociferous demands that authorities take forceful action against all the "old women", "Jews", "bunglers", and other uneducated healers[56] from whom many patients sought advice fell largely on deaf ears. This is not surprising. After all, those in charge themselves turned to such healers when they fell ill, and some physicians did admit that they knew of patients who experienced a noticeable improvement after having been treated by an

[54] Tellingly, we find this also in the letters addressed to Leonhard Thurneisser, personal physician to the Elector of Brandenburg. Thurneisser had never studied medicine but very successfully imitated the conduct of a learned physicians and was generously rewarded for his services; see, for example, Staatsbibliothek Berlin, Ms. germ. fol. 422b, letter from Antonius Lubbersdorff to Thurneisser, June 17, 1578.

[55] John Slater and Maria Luz López Terrada, "Scenes of mediation. Staging medicine in the Spanish interludes", in: *Social history of medicine* 24 (2011), pp. 226–243.

[56] See, for example, Ananius Horer, *Artzney-Teuffel, oder kurtzer Discurs, darinn diesem Ertzmörder seine Larve abgezogen*, sine loco 1634; see also Barbara Elkeles, "Medicus und Medikaster: Zum Konflikt zwischen akademischer und "empirischer" Medizin im 17. und frühen 18. Jahrhundert", in: *Medizinhistorisches Journal* 22 (1987), pp. 197–211.

uneducated healer. No less than the famous professor of medicine Vettore Trincavella in Padua told his students that old women with their herbal teas also could achieve remarkable results.[57]

Painful Compromises

In order to hold their own in the face of the stiff competition in the early modern health market, physicians, then, were perpetually forced to make unwelcome compromises that contravened their moral and scholarly principles, that chafed against their self-understanding as learned physicians. This found expression in various areas of physicians' medical practice.

"Mentiris ut medicus"—"you lie like a physician": this was a common saying at the time, and not without reason. A widespread piece of advice, in fact, was that the physician should always deceive his gravely ill patient until the end, whereas with the messenger or with relatives, he should couch the looming disaster in drastic terms. If, against expectations, the patient recovered, the physician would appear in an advantageous light, and if the patient did not, then at least the physician would have predicted the unfavorable outcome and it could be attributed not to his lacking abilities but to the severity of the illness. As Georg Handsch noted in 1550, those taking care of the patient could be told: "Things look truly bad for him, worrying, dangerous, he's hanging in the balance". The patient, however, one should "always console, not make him fainthearted, not abandon him".[58] Following this line of reasoning, when the condition of a fourteen-year-old febrile girl was increasingly deteriorating, he continued to give her parents hope. But to the maid he said that the sick girl would die. When she finally did succumb to her illness, the maid told the girl's parents that Handsch had predicted the death and, as he explicitly noted, "it pleased them".[59]

It was even difficult to consistently uphold recognized imperatives of medical deontology. It was considered every Christian physician's duty, for example, to care for the incurable and dying. In practice, however, many physicians patently avoided such patients as much as they could. In contemporary case histories, there is strikingly frequent mention of

[57] ÖNV, Cod. 11238, fol. 88r, note by Georg Handsch.

[58] ÖNV, Cod. 11206, fol. 100v.

[59] ÖNV, Cod. 11183, fol. 140r.

patients whose physicians had "given up" on them. Newcomers to the profession were advised not to let experienced physicians saddle them with these patients. The risk was too great that the poor outcome, inescapable as it was, would be attributed to with their incompetence, which would deal a severe blow to their reputation. Young Handsch noted down the advice of the famous Giovanni Manardo (1462–1536): "Ne suscipias morbos incurabiles, si famae tuae consultum esse cupis", which translates roughly as "do not treat incurable diseases if you care about your reputation". In his own practice, he did not always take heed of this advice—and bitterly regretted it. On various occasions, he reminded himself not to treat incurable patients again.[60]

Not least of all, the scientific principles of academic medicine as such were sometimes at stake. Uroscopy is a particularly illustrative example of how it was sometimes necessary for physicians to contravene their own ideas of rational medicine when confronted with the pressure of the marketplace.[61] In the scholarly literature of the Middle Ages, diagnosing illness on the basis of urine was a practice still extolled as the most important and reliable diagnostic procedure; it was the source of a physician's authority. In the early modern period, however, skepticism prevailed in the medical literature. The criticism was aimed only to a very limited degree at uroscopy as such, but it targeted the widespread expectation on the part of patients that physicians should be able to determine the diseases they suffered, whether or not they were pregnant, and indeed even their age and sex solely by looking at the urine and without any further information. For the critics, the risk of an embarrassing misdiagnosis or false prognosis was too great. How easily might it happen that a physician would diagnose a fatal illness only for the patient to be treated by a barber-surgeon or blacksmith and then be seen enjoying good health for years to come? How humiliating would it be if a physician were to diagnose a suppressed menstruation only for the woman to give birth to a child a few months later? It was even possible, the medical literature warned, that people purposefully made the physician a laughing stock by

[60] ÖNV, Cod. 11205, fol. 690v and fol. 691r; as Handsch noted here, his own teacher, Ulrich Lehner, also sought to avoid accepting incurable patients; in another notebook, Handsch repeated this advice to himself, mentioning a specific case (ÖNV, Cod. 11183, fol. 1r).

[61] On the following, see my detailed analysis in Michael Stolberg, *Die Harnschau. Eine Kultur- und Alltagsgeschichte*, Cologne and Weimar 2009.

slipping him cow's urine or even Malvasia wine. It was better after all to forgo uroscopy altogether. Yet physicians were faced with a problem. The less learned healers were very much willing to do the bidding of their patients and diagnose illnesses and pregnancies solely from the urine they were sent. And quite often, their results satisfied patients. To no avail, physicians sought to unmask the "tricks" used by their competitors, for example shrewdly questioning the urine messenger or listening behind a curtain when their wives sounded out the messenger. Ultimately, there was often no alternative for most physicians. Against their convictions, they too had to make their diagnoses predominantly, and sometimes exclusively, from the urine if they did not want to lose patients. They had to accept the fact that by doing this, they were more like those charlatans than they would have liked to admit, whose "fraud" they vociferously pilloried at every possible occasion.

From Learned Doctor to Experienced Physician

An option that was far more attractive than making such unpleasant compromises was to earn patients' trust by improving one's own diagnostic and therapeutic abilities and attaining superior curative success. Judged by today's standards, these efforts remained largely fruitless. They acted, however, as important catalysts for two very powerful processes in the learned medicine of the sixteenth century, which would permanently and profoundly change the public image of physicians as well as their self-understanding: (1) empirical experience was increasingly appreciated and (2) practical medicine and the knowledge and skills it required, increasingly came to be valued higher than theoretical, natural-philosophical knowledge.

In the learned medicine of the sixteenth and seventeenth centuries, the authoritative texts from antiquity and the Middle Ages remained a central point of reference, and the learned physician had to be well versed in them. At the same time, however, a belief in the importance of empirical experience too root in the medical community, even and especially among the humanist physicians. It was accompanied by a belief in progress that would be propelled by the innovative observations and ideas of individual physicians. In certain sub-fields such as the evaluation of healing springs, such a empirical toward the empirical can be traced back to the fifteenth century.[62] In the sixteenth century, this development became dramatically more dynamic. Botany, which was practiced first and foremost by physicians because of the paramount therapeutic importance of plants, became one of the most important topics in the *res publica medica*. The hope was that a better knowledge about plants and their medical properties would put the physicians' treatment of disease on a more secure footing.[63] Further, by actually dissecting corpses rather than only reading the works of Galen, physicians sought to discover and unlock the mysterious pathological processes within the body. But above all, many physicians were now attempting to improve their diagnoses and therapies by observing patients and their illnesses closely at the bedside. Some of them made detailed notes in their practice diaries and notebooks of the cases they were treating and the effects of different medications. For, according to a growing conviction, the active properties of drugs could not be deduced solely from theoretical assumptions or from the sensorily perceptible qualities of the plants in question—warm, cold, dry, and moist. Rather, their action on diseases was often of an "occult", specific nature and could only be deduced from the effects one observed in the patient.[64]

In physicians' applications for medical positions, we find that references to knowledge of special, proven remedies—those held to be particularly effective—sometimes took a prominent place. In 1599, Matthaeus Gloning, for example, claimed that he had not only successfully worked for all manner of noble gentlemen, but had also "collected many excellent things that I am confident, with God's help, I may use for the special benefit of my patients in appropriate cases".[65] A minority of physicians took this one step further. Some attempted to isolate and concentrate

[62] Katherine Park, "Natural particulars: Medical epistemology, practice, and the literature of healing springs", in: *Natural particulars. Nature and the disciplines in Renaissance Europe*, ed. by Anthony Grafton and Nancy Siraisi, Cambridge and London 1999, pp. 347–367.

[63] Some of the most influential works were Leonhard Fuchs, *De historia stirpium commentarii insignes*, Lyon 1547; Pierandrea Mattioli, *Commentarii in sex libros sex Pedacij Dioscoridis Anazarbei de medica materia*, Venice 1570; early already, some vernacular works were published as well; see, for example, Otto Brunfels, *Contrafayt Kreüterbuch*, Strasbourg 1532.

[64] Michael Stolberg, "Empiricism in sixteenth-century medical practice. The notebooks of Georg Handsch", in: *Early science and medicine* 18 (2013), pp. 487–516.

[65] Stadtarchiv Augsburg, Collegium medicum, Karton 4; not dated but the town council decided on his request on April 3, 1599.

the active ingredient of a medicine in their laboratories in the form of a quintessence. Others even carried out experiments, testing, for example, the effectiveness of antidotes against aconite and other poisons on men condemned to death.[66] Here, we can discern a long-term shift away from the learned, well-read natural philosopher to the empirical "natural scientist". This shift began taking place in medicine well before the empirical program of Francis Bacon and well before the much-invoked "scientific revolution" of the seventeenth century—in which medical practitioners had a decisive role, as recent research has brought to light.[67]

The turn toward empiricism, toward personal observation as a central source of knowledge, went hand in hand with a growing appreciation for the concrete practical knowledge and specific skills that the physician needed if he wanted to stay afloat amid the great competition in the contemporary medical marketplace. This growing appreciation of *medicina practica* was also reflected in teaching, beginning with the northern Italian universities, the leading institutions in Europe, where many future physicians from north of the Alps came to study. Alongside lectures on authoritative texts, practical instruction at the bedside—in hospitals or accompanying the professors when they made house calls to patients—became an indispensable part of medical education and was very much appreciated by the students: they were trained in performing uroscopy and feeling the pulse, and they learned how to ascertain on the basis of close observation at the bedside the presumptive internal cause of disease and the therapies that could address this cause.[68]

Even manual skills and procedures, which physicians had hitherto gladly left to the barber-surgeons, gained increasing relevance, both in the university context and in everyday medical practice. As early as the sixteenth century, many physicians—and this has been almost completely ignored in historical research so far—palpated their patients and this skill,

[66] ÖNV, Cod. 11183, foll. 125r-127r, December 11, 1561 [?]; cf. Pierandrea Mattioli, *New Kreutterbuch mit den allerschönsten vnd artlichsten Figuren aller Gewechsz, dergleichen vormals in keiner Sprach nie an Tag kommen*, Venice 1563, foll. 472r-v.

[67] Harold J. Cook, "Victories for empiricism, failures for theory. Medicine and science in the seventeenth century", in: *The body as object and instrument of knowledge. Embodied empiricism and early modern science*, ed. by Charles T. Wolfe and Ofer Gal, Dordrecht 2010, pp. 9–32.

[68] Michael Stolberg, "Bedside teaching and the acquisition of practical skills in mid-sixteenth-century Padua", in: *Journal of the history of medicine and allied sciences* 69 (2014), pp. 633–661.

too, was taught at some universities as early as the sixteenth century. They examined them for tumors and indurations, inserted probes, and even personally performed gynecological examinations.[69]

The appreciation for sensorial and manual skills and practical experience at the sickbed can also be observed in the application letters which physicians wrote to municipal authorities.[70] It was not their erudition but the practical experience they had acquired—and of course their curative successes—that constituted the golden ticket with which they sought to gain the position of town physician or a position at a municipal hospital. In 1549, the Wasserburg physician Georg Haindlacher explained, for example, that in his fourteen years of practice, he had "come to such practice and experience and [had] worked with such diligence" that he easily was a match for other physicians, by which he meant those in Augsburg in particular.[71] Likewise extolling his experience in 1606, Jacob Berckhmüller wrote that he not only had studied at different universities, "but also practiced in the medical field at the renowned and praiseworthy Hospital Santa Maria Nuova in Florence and in other places to such an extent that now, with divine assistance, I have cause to apply my talentum".[72] Urban Schlegel for his part, pointed to the practical experience he had gained in hospitals in Munich.[73] Other physicians shone a light on the experience they had already gained in university. Johann Conrad Zwilling, for example, wrote that he had "gained practical exercise under the direction of physicians for two to two and a half years [...], in addition to attending the collegia".[74] In their letters of recommendation,

[69] Michael Stolberg, "Examining the body (c. 1500–1750) ", in: *The Routledge history of sex and the body, 1500 to the present*, ed. by Sarah Toulalan and Kate Fisher, Oxford 2013, pp. 91–105.

[70] See Sabine Schlegelmilch, "Promoting a good physician. Letters of application to German town authorities (1500–1700)", in: *Civic medicine. Physician, polity and pen in early modern Europe*, ed. by Andrew Mendelsohn, Annemarie Kinzelbach, and Ruth Schilling, Abingdon 2020, pp. 88–109.

[71] Stadtarchiv Augsburg, Collegium Medicum, Karton 4, letter from Haindlacher, 1549.

[72] Ibid., letter from Berckhmüller, 1606.

[73] Ibid., letter from Schlegel, November 15, 1618.

[74] Stadtarchiv Lindau 49,6, undated letter, received on December 20, 1697.

physicians' previous employers likewise praised them with words like "an experienced and lauded medicus".[75]

This shift in physicians' self-understanding and in the external perception of learned physicians carried a considerable risk. With the new emphasis on experience and practical skills, the image of the academically trained physician was drawing perilously near to that of the craftsman. And worse yet, the physician, who no longer relied on theory to determine the choice of medicines for certain diseases, but instead on his observation of their effects on patients with similar symptoms, seemed dangerously akin to the uneducated "bungler", the *empiricus* with his supposedly tried and tested secret remedies. In the long term, however, this embracing of more empirical and practical approaches was successful. Among the wealthy, educated classes on whom physicians depended first and foremost for their prosperity and professional advancement, the physicians' academic education and the cultural horizon they shared with these patients apparently secured sufficient trust to allow the physicians to prosper. Medical self-fashioning—both the self-understanding of physicians and their self-staging to the outside world—began to follow a new model: physicians were now not only academically trained but more importantly, they were seasoned practitioners who took pride in their tried and tested experience.

[75] Stadtarchiv Augsburg, Collegium Medicum, Karton 4, letter of recommendation by Palatine Count Wolfgang Wilhelm zu Neuburg, 1615.

Forms and Functions of Medical Case Reports in the Early Modern Period (1500–1800)

Case reports have played an important role in medicine since antiquity.[1] This esteem for the study of individual cases was founded in no small way on the fact that medicine was not only a theoretical but a practical discipline. It sought not only to gain general knowledge about the nature of man and his diseases, but it also aimed—even in the eyes of learned, philosophical physicians—to apply this knowledge successfully to individual cases of illness.

The meaning and function of medical case reports repeatedly changed markedly, however, in the course of time, reflecting, in particular, the relative importance attributed to practice and experience vis-à-vis the trust that was placed in authoritative texts or overarching medical theories.

First published in German as "Formen und Funktionen ärztlicher Fallbeobachtungen in der Frühen Neuzeit (1500–1800)", in: *Fallstudien: Theorie – Geschichte – Methode*, ed. by Johannes Süßmann, Susanne Scholz, and Gisela Engel, Berlin: trafo-Verlag 2007, pp. 81–95. Translated into American English by Logan Kennedy and Leonhard Unglaub.

[1] See Hippokrates, *Epidemics*. I am grateful to the participants of the Frankfurt conference organized by the editors and to Peer Otte for their valuable feedback on previous versions of this paper.

M. Stolberg, *Medicine and the Body in Early Modern Europe*, Palgrave Studies in Medieval and Early Modern Medicine,
https://doi.org/10.1007/978-3-031-84556-7_8

The early modern period, which is the focus of this article, constitutes a high point in Western medical history with respect to the value physicians placed on case studies. Beginning in the late sixteenth century, numerous physicians, many of them leading authorities of their day, published case histories from their own practice or compiled case reports that colleagues had communicated to them orally or in writing or else which they found in older[2] or more recent[3] literature. With the emergence of scientific journals at the end of the seventeenth century, a large number of personal medical observations also came to be published in periodicals. This tradition of publishing one or several similar cases would then be carried on and intensified in the medicine of the nineteenth century, which is beyond the scope of this article.

I will primarily concentrate here on the medical-literary genre of medical case report collections as opposed to individual case reports. In recent times, medical historians have discovered how valuable these collections are, in particular with respect to the social and cultural history of medicine. Medical case reports allow today's reader to gain manifold insights into contemporary medical life and into the medical ideas and practices of physicians and laypeople.[4] The historical development of the genre, however, and its various forms and functions, have not been sufficiently investigated so far.[5]

[2] Symphorien Champier, *Claudii Galeni Pergameni historiales campi [...] in quatuor libros congesti et commentarijs non poenitendis illustrati*, Basel 1532.

[3] A prominent example is Johannes Schenck, von Grafenberg, *Observationum medicarum, rararum, novarum, admirabilium, et monstrosarum tomus unus*, Frankfurt am Main 1600.

[4] Barbara Duden, *Geschichte unter der Haut. Ein Eisenacher Arzt und seine Patientinnen um 1730*, Stuttgart 1987; Katharina Huber, *Felix Platters 'Observationes'. Studien zum frühneuzeitlichen Gesundheitswesen in Basel*, Basel 2003; Karen Nolte, "Vom Verschwinden der Laienperspektive aus der Krankengeschichte. Medizinische Fallberichte im langen 19. Jahrhundert", in: *Zum Fall machen, zum Fall werden. Wissensproduktion und Patientenerfahrung in Medizin und Psychiatrie des 19. und 20. Jahrhunderts*, ed. by Sibylle Brändli-Blumenbach, Barbara Lüthi, and Gregor Spuhler, Frankfurt and New York 2009, pp. 33–61.

[5] Pedro Laìn Entralgo, *La historica clinica. Historia y teoria del relato patografico*, Barcelona 1961, remains an important point of departure but in many respects calls for further analysis; further valuable information in Gianna Pomata, "*Praxis historialis*. The uses of *historia* in early modern medicine", in: *Historia. Empiricism and erudition in early modern Europe*, ed. by eadem and Nancy G. Siraisi, Cambridge, MA, 2005, pp. 105–146; Vivian Nutton, *Case-histories in the early Renaissance*, unpublished lecture

In some respects, the great significance of case histories in early modern medicine derives from a more comprehensive revaluation of *historia*, that is, in an Aristotelean sense, non-demonstrative knowledge as opposed to the logical and rational deductive knowledge in contemporary natural philosophy as a whole.[6] But medical case reports, in particular due to their close connection to medical practice, also have some unique characteristics. With this in mind, the aim of this paper is to bring into focus the essential characteristics and the forms and functions of medical case reports in the early modern period.

Consilia

Already toward the end of the Middle Ages, the *consilium* emerged as a relatively clearly demarcated casuistic genre in its own right. A *consilium* was a letter of medical advice for a particular patient. The wealthy ill or their physicians would quite frequently request such counsel, in particular if the illness was protracted or if the treatment undertaken sofar had not yielded satisfactory results. The author of a *consilium* would respond by explaining, generally at length, the presumed pathophysiological processes within the body and establishing his diagnosis and prognosis on this basis. From this, he could proceed to his therapeutic advice, which generally included detailed dietetic recommendations concerning nutrition and lifestyle.

By the late Middle Ages, the *consilia* of famous physicians such as Taddeo de' Alderotti, Ugo da Benzi, and Gentile da Foligno had been collated into larger handwritten collections, and many copies of them were in circulation.[7] In the sixteenth and seventeenth centuries, some of these collections of *consilia* were then published in print along with newer collections by living authors.[8] These collections did more than

manuscript, Stuttgart 1991; Johanna Geyer-Kordesch, "Medizinische Fallbeschreibungen und ihre Bedeutung in der Wissensreform des 17. und 18. Jahrhunderts", in: *Medizin, Gesellschaft und Geschichte* 9 (1990), pp. 7 19.

[6] Pomata and Siraisi, Historia [see note 5].

[7] See the in-depth analysis in Jole Agrimi and Chiara Crisciani, *Consilia médicaux*, Turnhout 1994; cf. also D. P. Lockwood, *Ugo Benzi. Medical philosopher and physician, 1376–1439*, Chicago 1951.

[8] Giovanni Battista da Monte, *Consilia medica omnia, quae ullibi extant*. Ed. by Girolamo Donzellini, Nürnberg 1559; Vettore Trincavella, *Consilia medica*, Basel 1587;

teach students and budding physicians the (financially very rewarding) art of penning *consilia*, sometimes accompanied by general and systematic instructions on how to write them.[9] They also taught the reader how to draw reliable conclusions about the individual case with its many particularities based on generalized medical concepts, thus arriving at astute diagnoses and a promising treatment—comparable with those enlightening bedside lessons for which Giovanni Battista da Monte and other leading contemporary Italian professors of medicine were famous.

Ordinary Cases in Medical Practice: Observationes, Curationes, Casus

In the eighteenth century, famous physicians such as Lorenz Heister and Samuel Tissot continued to author numerous *consilia* for patients from far-flung places. *Consilia* as a literary genre were, however, increasingly displaced by collections of retrospective medical case reports that described the illness from its beginnings all the way to its happy or occasionally bitter end.[10] Only forensic *consilia* maintained their significance and indeed became more meaningful. These were written reports delivered by individual physicians or entire medical faculties in their increasingly valued role as experts before the court.[11]

Using titles such as *observationes* and *curationes medicinales* or, more rarely, *historiae* or *casus*, practice-based medical case reports, beginning in the late sixteenth century, rapidly developed into one of the most popular

Ioannis Baverius, *Consiliorum de re medica sive morborum curationibus liber*. Ed. by Walther H. Ryff, Strasbourg 1543.

[9] Benedictus Silvaticus, "De consultandi ratione", in: idem, *Consiliorum et responsorum medicinalium centuriae quatuor*, pp. 610–620; Roderigo de Fonseca, *Consultationes medicae singularibus remediis refertae*, Frankfurt 1620, pp. 309–322.

[10] See, however, Friedrich Hoffmann, *Medicina consultatoria*, 12 vols, Halle 1721–1739; Joh. Andreas Fischer, *Consilia medica iterum continuata, quae in usum practicum et forensem pro scopo curandi et renunciandi adornata atque singularibus experimentis, inter arcana domestica hactenus reservatis, illustrata sunt*, Frankfurt am Main 1712.

[11] See, for example, Michael Bernhard Valentini, *Pandectae medico-legales, sive responsa medico-forensa*, Frankfurt 1701; on the genre, in general, Maren Lorenz, *Kriminelle Körper - gestörte Gemüter. Die Normierung des Individuums in Gerichtsmedizin und Psychiatrie der Aufklärung*, Hamburg 1999.

and most frequently cited genres of early modern medical literature.[12] Above all, physicians in the wealthy imperial cities and trading centers north of the Alps distinguished themselves with extensive collections of case histories taken from their own practice, sometimes comprising numerous volumes. Some, like Pieter van Foreest in Alkmaar and Felix Platter in Basel,[13] were in the employ of a municipality. Others—Tulpius in Amsterdam, Heer in Lüttich, Hechstetter in Augsburg, and Throner in Ulm, for instance—ran an extensive private practice.[14] Medical practice in sizeable cities offered the best conditions for assembling these kinds of medical case reports. Unlike the court physicians and the professors of medicine in smaller university towns such as Wittenberg and Helmstedt, physicians in major cities often cared for large numbers of patients, including less well-to-do patients and the poor sick. Sometimes they were also in charge of medical care in hospitals and similar care facilities. They were, in other words, able to draw on their personal experience treating hundreds or thousands of patients.

Unlike with *consilia*, medical case reports did not arise directly and immediately from the physicians' everyday practice. Some authors seem to have followed the advice of their teachers and systematically noted down practical experiences and thus acquired a constantly growing body of empirical knowledge. Tancredus Lelius, for example, explicitly characterized his twenty-five *curationes* as "exerptae ex diario practico".[15] Other physicians likely had their eye on a potential future publication when they

[12] François Valleriola, *Observationum medicinalium libri sex*, Lyon 1573; Rembertus Dodonaeus, *Medicinalium obseruationum exempla rara, recognita et aucta*, 2nd edn, Cologne 1581; Pieter van Foreest, *Observationum et curationum medicinalium libri XXXII*, Leiden 1587–1610.

[13] Foreest, Observationum [see note 12]; Felix Platter, *Observationes in hominis affectibus plerisque corpore et animo [...] incommodantibus libri tres*, Basel 1614.

[14] Nicolaus Tulpius, *Observationum medicarum libri tres*, Amsterdam 1641. Philipp Hechstetter, *Rararum observationum medicinalium decades tres. Continentes historias medicas, theorica et practica varia, iocunda, utilia, necessaria*, Augsburg 1624; Henricus ab Heer, *Observationes medicae oppido rarae in Spa & Leodii animadversae, cum medicamentis aliquot selectis, & ut volunt secretis*, 2nd edn. Ed. by Joannis Michaelis, Leipzig 1645; Augustin Thoner, *Observationum medicinalium haud trivialium libri quatuor*, Ulm 1651.

[15] Edited in Thomas Bartholin, *Cista medica hafniensis*, Copenhagen 1657, p. 409; see also Heinrich Meibom's preface to Arnoldus Bootius, *Observationes medicae de affectibus omissis*, 2nd edn, Helmstedt 1664.

set out to take notes, or at least they thought of sending their observations by letter to a respected colleague—a common form of "publishing" at the time.

Like the *consilia*, collections of medical case reports from everyday practice generally covered the whole spectrum of diseases. However, physicians usually selected the more important, interesting, and particularly illuminating cases—and preferably those with happy endings. The latter had the unmistakable advantage of casting their diagnostic and therapeutic abilities in a very positive light: medical case histories always also played a role in self-representation. They portrayed their authors as particularly skilled diagnosticians and therapists, thus also drawing a clear line between them and their less-educated competitors who were often looked on very highly by patients. The allegedly disastrous misjudgments of these practitioners were usually described at length and with gusto. For learned surgeons such as Wilhelm Fabry, this kind of self-representation even played a crucial role. Theirs was a primarily manual form of medicine—quite akin to the work of craftsmen—and for them, the competition from less-educated surgeons was especially challenging. With medical case reports, these educated surgeons were able to highlight the superiority of their refined theoretical knowledge and their outstanding grasp of anatomy and to contrast it to the devastating results of competing barber-surgeons and "bunglers", who had possibly tried their hand at the very same case before or after them.[16]

Of course, it was also in the interest of the reader to have a selection of particularly difficult cases of illness that were treated with success. In the contemporary understanding, learning *per exempla* was an indispensable complement to learning *per praecepta*.[17] For this reason, the importance of reading collections of medical case studies was impressed upon students of medicine.[18] Collections of difficult cases added to the limited personal experience of the individual physician in an essential, indispensable way. They familiarized him with clinical pictures that only seldom appeared in his own practice. And they showed him through concrete examples

[16] Wilhelm Fabricius, *Observationum & curationum chirurgicarum centuriae*, Basel 1606–1641; Johannes Scultetus, *Wund-Artzneyisches Zeug-Hauß*, Frankfurt 1666, with a separately paginated collection of *observationes*.

[17] Meibom, Preface [see note 15].

[18] See, for example Johannes Rhodius, "De artis medicae exercitatione, consilia tria", in Bartholin, Cista [see note 15], pp. 616–629.

how he, by looking at the medical history and the clinical picture, could deduce the nature and the primary seat of even complicated diseases, how he could name the possible *causae proximae* and *remotae*, and finally, how he could formulate his prognostic judgment and his therapeutic advice.[19]

The length and the content of such practice-based medical case reports varied greatly. Some focused almost exclusively on therapeutic aspects, which was sometimes already expressed in the titles, such as *curationes* and *curationes difficiliores*. They linked back to the genre of *curationes*, which Amatus Lusitanus first and foremost made popular around 1560.[20] These were reports of successful treatments which were often presented in no particular order, simply one after the other. Another line of tradition leads to the numerous collections of tried and tested medications found in handwritten and printed recipe books, which pointed out certain medications for certain clinical pictures, not infrequently with reference to their successful application in illness of a specific, named patient. In such primarily therapeutic, curative medical case reports, the causal interpretation of the medical condition was of secondary importance or could even only implied by reading between the lines. Above all, these works provided the medical reader with a tested therapeutic arsenal. Now all he had to do was adapt the remedies that were considered effective to the unique constitution of the respective patient. Other medical case reports were largely limited to diagnosis, to identifying an unusual medical condition, and they mentioned treatment only peripherally. This was especially true of clinical pictures whose interpretation demanded special diagnostic acumen.

As a supplement to medical case reports, pathological-anatomical findings came to be increasingly valued. As early as the sixteenth and seventeenth centuries, they were occasionally integrated into the text or added as a separate appendix. Postmortem findings offered a welcome, seemingly objective means of establishing in retrospect whether or not the diagnosis or treatment of a certain physician—or his competitors—had been correct. Postmortem findings, which shed light on the massive pathological changes in the body, which even the best therapy could not

[19] A good example is Michael Ettmüller, "Collegium consultatorium", in: idem, *Opera omnia theoretica et practica*, Lyon 1685, pp. 141–212.

[20] Amatus Lusitanus, *Curationum medicinalium centuria prima, multiplici variaque rerum cognitione referta*, Florence 1551.

put a halt to, also made it easier to describe an unfavorable progression of an illness without damaging one's own reputation.

Over time, collections of autopsy findings developed into an independent casuistic genre. Early in the sixteenth century, Antonio Benivieni published a first major collection of medical case reports, many of which were appended by a postmortem report.[21] In the late seventeenth century, Théophile Bonet's *Sepulchretum* assembled numerous postmortem findings from the writings of older and more recent authors.[22] In the eighteenth century, Giovanni Battista Morgagni led the way in directing medical attention to smaller, less obvious changes that occurred in different organs with different diseases. Initially sharing his numerous observations with other physicians by letter, he eventually, in 1761, collected them in his pioneering *De sedibus et causis morborumper anatomen indagatis libri V*.[23] The connection that was being made here between clinical picture and pathological finding would eventually prove to have a guiding function for modern clinical medicine. With the help of the postmortem pathological finding, seemingly similar clinical pictures could be distinguished from one another and, if necessary, attributed to different pathophysiological processes. For centuries, for example, dropsy was considered a discrete illness until dissections revealed that it could stem from a hardened liver, an enlarged heart, or small, hard kidneys. Collections of case histories with postmortem findings thus signal a transition from medical case reports that primarily had an applied function and aimed at practical instruction toward forms of medical case studies that were meant to go beyond the systematic collection of individual observations and instead expand medical knowledge as a whole.

[21] Antonio Benivieni, *De abditis non nullis ac mirandis morborum et sanationum causis*, Florence 1507.

[22] Théophile Bonet, *Sepulchretum anatomicum seu anatomia practica ex cadaveribus morbo denatis proponens historias et observationes, quae pathologiae genuinae tum nosologiae orthodoxae fundatrix dici meritur*, Geneva 1679.

[23] Giovanni Battista Morgagni, *De sedibus et causis morborum per anatomen indagatis libri V*, 2 vols, Venice 1761.

Natural Historical Case Reports: Observationes Rarae or Curiosae

There were no clearly demarcated boundaries between medical case reports that described rare illnesses, which the physician might only have learned about second hand, and reports that stood in the natural historical tradition of collecting observations of singular, very unusual phenomena. The latter described grotesque deformities, peculiar medical symptoms, rare bodily phenomena, and in fact any and all things bizarre from the whole kingdom of nature, which appeared inexplicable, at first glance at least, perhaps suggesting a supernatural origin.[24] Hellwig, for example, reported the case of a pregnant woman who had drunk from puddles. Soon, she heard frogs croaking in her belly and felt strange sensations inside her body. Eventually she took some remedies, whereupon she vomited two live frogs and discharged two further, dead frogs with her stool.[25]

Natural historical observations and case reports showed above all the near infinite multitude of phenomena that nature was able to bring forth. Thus, they not least satisfied the curiosity and desire for sensation of authors and readers alike. Like the owners of curious artifacts, some authors were unabashedly proud of their particularly original observations. However, stories of extraordinary cases of illness and other rare occurrences also posed challenges, which were at times quite welcome, for the theoretical discourse. After all, it was in dealing with the seemingly inexplicable, with apparent contradictions and anomalies, that the explanatory power of a theory could be demonstrated. For the common layman, it did not seem in line with the laws of nature for a woman to have living frogs in her body, but the learned physician, as Hellwig wrote, "must not immediately consider everything to be a fable that seems so at first glance".[26] This was true even in cases where something that was observed seemed at first glance to contravene the laws of the human body. After all, according to the widespread credo of contemporary physicians and natural philosophers, everything in the world obeyed the laws

[24] In the case of Marcello Donato, *De medica historia mirabili libri sex*, Venice 1588, the title itself already clear states the focus.

[25] L. Christoph Hellwig, *Casus et observationes medicinales, anatomicae, chymicae, chirurgicae, physicae, etc. rariores, selectae, et curiosae*, Frankfurt and Leipzig 1710, p. 53.

[26] Ibid.

that God laid down when He created the world. In the final assessment, then, even the most peculiar occurrences had to be explained as unfolding within the natural order, and if necessary, the dominant theories had to be adapted, refined, modified, or supplemented accordingly.[27] If a woman gave birth to frogs, learned physicians could point to the widely accepted possibility of spontaneous generation, or else they could interpret the occurrence the way Hellwig did as proof of the undiminished potency of frog seed (in the puddle water), which could develop into a grown frog even in the body of a woman.

Such *observationes rarae* were, however, hardly of any relevance to the practitioner's work. Jakov van Hervelt, for example, in his *Geneeskundige aanmerkingen*, explicitly distanced himself from them. His readers, he emphasized, were not to expect stories of miraculous events, strange monstrosities, and the like. Such things only happened every century or so and thus their description was of little or no practical use. As a writer of practice-based medical case reports, he instead wanted to focus entirely on the very quotidian cases.[28]

Given the almost unbelievable diversity of observed nature, a systematic collection of the greatest number of unusual case histories and indeed any kind of factual observations was most readily realizable in the form of a collective project, either in the public realm of scholarly medical publications or, in a more structured form, as part of the activities of an academy. The *Ephemerides* of the Leopoldina offer an illustrative example here. The academy's members, who in the beginning were all medical professionals, worked together to collect noteworthy observations of every kind and to publish them in sequence, according to the they were submitted.[29] Similar efforts were made outside formal academies as well. For years, the Nuremberg journal *Commercium litterarium*, for example, collated personal observations sent in by a network of corresponding physicians.[30]

[27] Lorraine Daston and Katharine Park, *Wonders and the order of nature* 1150–1750, New York 1998.

[28] Jacov van Hervelt, *Geneeskundige aanmerkingen in sijn dagelijke practyk voorgevallen*, Amsterdam 1693, preface to the reader.

[29] *Miscellanea curiosa sive ephemeridum medico-physicarum Germanicarum Academiae Caesareo-Leopoldinae Naturae Curiosorum*, 1670–1706.

[30] *Commercium litterarium ad rei et scientiae naturalis incrementum institutum quo quicquid novissime observatum agitatum scriptum vel peractum est succincte dilucideque exponitur*, Nürnberg 1731–1745.

Publishing through a recognized scientific institution was at the same time an effective answer to a problem that the authors of natural historical observations struggled with more than physicians who reported from their own practice, namely the issue of credibility. When looking at reports of strange, sensational occurrences or reports of the amazing recoveries of patients with fatal diseases, the contemporary reader always had to wonder whether he could truly believe the author. Perhaps a certain degree of trust was established by the author's good reputation. For their part, the authors sought to forestall possible doubt, especially by providing detailed information about the patient in question, such as his or her profession, place of residence, family affiliation, and sometimes further information—at the time, medical confidentiality was only recognized to a very limited extent. In this way, they communicated clearly to their readers that they were presenting authentic, verifiable observations rather than something fictitious. Serving a similar function were illustrations that were true to detail, for example in the case of deformities. Even the new ideal of observation in the presence of credible witnesses, which recent research in the history of science has highlighted as a preeminent innovation of the Royal Society,[31] had been in the making decades earlier in some medical case reports, which named the physicians or other educated witnesses who were present and could confirm the report and put the doubting reader's mind at ease.

Case History and Theory

On the face of it, the medical case histories of the early modern period constituted a decidedly empirical genre. Initially, in the late sixteenth century, their popularity stemmed not least of all from a growing frustration with the dominant Galenic medicine, which came to be perceived as theory-heavy, speculative, and self-contradictory. Its gradual loss of significance in the seventeenth and eighteenth centuries naturally fed the skepticism about theoretical approaches. In the subsequent time, as different medical currents came to exist in parallel and in opposition to one another—from Helmontian iatrochemistry to mechanical-hydraulic and vitalist conceptions all the way to Stahlianism and Brownianism—the

[31] Steven Shapin, *A social history of truth. Civility and science in seventeenth century England*, Chicago and London 1995.

common bedrock of medicine was removed, with no new and generally recognized theoretical foundation that was put in its place. In a situation in which, as lamented by Friedrich Hoffmann, "almost every famous professor wants to formulate a special theory",[32] it seemed to many authors that a return to observation as unobstructed by prejudice as possible was the best way forward. Giorgio Baglivi, for example, wrote that scorn for the old physicians, a love of systems, preconceived opinions, and incorrect theories had led medicine astray. According to him, it was important to collect observations and also to "to frequently visit the dirty sickbeds, the public lazarettos and hospitals" so as to carefully document what one sees there and, in particular, which "kind of treatment most constantly had a positive effect with each different disease".[33]

With their concise, to-the-point form, natural historical case reports clearly underscored their aim of communicating only objective facts, independently of any medical or philosophical system to which the author subscribed.[34] It was only in a second step that the case histories could become the rich soil in which theoretical reflection could grow. The intensive early modern discussion about the nature of female menstruation offers an illustrative example here. A plausible theory of menstruation had to take into account the case histories documenting the regular, sometimes even monthly bleeding observed in men—from hemorrhoids, from the nose, from the bowels, etc. And indeed, the prevalent "hydraulic" doctrine of menstruation of the seventeenth and early eighteenth centuries offered a plausible explanation for both female and male "menstruation". In the bodies of women—and sometimes also of men, especially if they ate a lot and did not move much—blood was always accumulating and stretching the vessels until they burst in the places where they were weakest. With women, it was the vessels of the uterus that were weak as a rule, while with men it was the hemorrhoidal veins. When, in the eighteenth century, the attempt was made to explain menstruation as a secretion of a gland in the uterus, it became necessary to find an explanation for male "menstruation"—or cast great doubt

32 Friedrich Hoffmann, *Medicina consultatoria*, 12vols, Halle 1721–1739, preface to vol. 3.

33 Giorgio Baglivi, *Zwey Bücher de praxi medica*, Lübeck and Frankfurt 1705, preface and p. 42 (cit.).

34 Lorraine Daston, "Perché i fatti sono brevi?", in: *Quaderni storici* 108 (2001), pp. 745–770.

upon the relevant case histories. After all, men certainly did not have a uterus and thus could not have a uterine secretion.[35]

As opposed to these natural historical case histories, the case histories from everyday medical practice were often more clearly framed within a theory, even though most of their authors made a point of following a purely descriptive, empirical model, and some of them even explicitly invoked Hippocrates; with his succinct case histories in the *Epidemics*—they were often only a few lines long—he had become the paragon of empirical medicine. The English physician Thomas Sydenham even gained a reputation as a second Hippocrates by aiming to simply observe the course of different diseases so as to determine, much in the way of botanists, the different species of disease.[36]

Upon closer observation, however, the Hippocratic case histories already prove to be saturated with theory, for example when they limit themselves to the description of the excretions on so-called "critical" days. Many practice-based case reports of the early modern period gave theory even more room. They tended to follow the Galenic ideal of a physician with a (natural-)philosophical education who relied on *experientia* and *ratio* equally. Just as with the older *consilia*, early modern practice-based case histories seldom lacked references to the presumed pathophysiological background of a given disease and to the relevant passages in the texts of ancient or contemporary authorities. These references could be made in the text itself or take the form of learned *scholia*.[37] The necessity for such theoretical explanations of medical conditions resulted from the didactic function of these case histories. It was the ambition of learned medicine—setting itself off from the world of the uneducated *empirici*—to treat diseases in a causal manner, that is to eradicate them or, literally, to pull them out by their roots. This required physicians to learn to take the clinical picture and the patient's history into account to identify the presumed disease causes within the body, which would then be the target of their therapy—as good case histories taught. Drawing on scholarly

[35] Cf. Michael Stolberg, "Erfahrungen und Deutungen der weiblichen Monatsblutung in der Frühen Neuzeit", in: *Scientiae et artes. Die Vermittlung alten und neuen Wissens in Literatur, Kunst und Musik*, ed. by Barbara Mahlmann-Bauer, Wolfenbüttel 2004, pp. 913–931.

[36] Thomas Sydenham, *The whole works*, 6th edn. Corrected from the orig. Latin by John Pechey, London 1715.

[37] Foreest, Observationum [see note 12].

pathophysiological explanations and sometimes by making reference to passages in authoritative texts, physicians in their everyday practice subsequently worked to convince patients and their relatives of the necessity and superiority of their prescriptions.

In some case reports, theoretical discussion took up so much room that the case at hand, as with Galen himself,[38] served only as a segue or an illustrative example. Some publications carrying the title *observationes* did not even offer any actual case histories.[39] Rather, they were treatises on particular diseases or medical subjects; in English, the word "observation" carries this extended meaning to this day. With respect to their genre, such works are much closer to the early modern *epistolae medicinales*[40] and to the subsequent scientific article on a specific medical subject than they are to medical case studies.

Conclusion

Upon close consideration, the rise of medical case studies in the medical writing of the early modern period amounts to a history of quite different text types, each composed in a different way and with different intentions. Their thematic spectrum extends from cases of illness in everyday medical practice all the way to more or less unique occurrences that happened as if by magic. Their form ranges from succinct reports of factual observations to learned treatises in which the specific empirical observation of the case served only as a segue or illustrative accessory.

There is no mistaking, however, that amidst all these differences, the case reports of the early modern period, beginning in the late sixteenth century, assumed a preeminent role as both an oft-cited genre and as empirical building blocks that could be used to found, call into question, or refine theories. In their richness, significance, and broad impact, they raise the question once again whether medicine must be understood—much more than it has been so far—as an important driving force behind Baconian empiricism and the so-called scientific revolution of the

[38] Cf. Champier, Claudii Galeni Pergameni historiales campi [see note 2].

[39] Iodocus Lommius, *Medicinalium observationum libri tres*, Antwerp 1560; Johann Weyer, *Medicarum observationum rararum liber I*, Basel 1567; Johann Nicolas Pechlin, *Observationum physico-medicarum libri tres*, Hamburg 1691.

[40] Johannes Lange, *Epistolarum medicinalium volumen tripartitum*, Hannover 1605.

seventeenth century. This is true not only with respect to the involvement of physicians in the collection of natural historical observations, but also—and importantly—with respect to practical medicine with its great significance in everyday life. Francis Bacon characterized medicine as an *ars coniecturalis*. Its object, the human body, was highly variable in its shape, highly subject to change, and was constructed in a more complex manner than any other natural body.[41] This implied that the results of medical activity could be predicted to a very limited extent and almost never reliably. Only empirical experience could tell, case after case, whether the physician had made the right diagnosis and had deduced the correct treatment from his theoretical knowledge. Thus, only empirical experience promised to help improve the methodical application of theoretical concepts to specific cases. Conversely, however, it was only through a theoretical understanding of case studies that academically trained physicians could arrive at the medical knowledge they needed to stand their ground at the bedside and to distinguish themselves from the countless less-educated competitors, the *empirici* above all, who administered ostensibly tried and tested remedies without carefully weighing the specifics of a given case of illness and without making the respective allowances as demanded by the art of medicine.

[41] Francis Bacon, *De dignitate et augmentis scientiarum libri IX*, Paris 1624, pp. 211–213.

Casuistic Medical Training in the Sixteenth Century: The Paduan *Collegia*

In the sixteenth and seventeenth centuries, medical training at universities saw profound changes. Whereas studying and commenting on the great authorities of antiquity and the Middle Ages was formerly the focal point, now practical training was gaining significance.[1] An important driving force of this development was the growing appreciation among the upper classes for education and erudition in general. In the course of the sixteenth century, it was not only princes and noblemen, but also patients from the middle classes who increasingly preferred, at least with regard to internal diseases, learned physicians over barber-surgeons and bath-masters, who were trained as craftsmen, and the countless male and female

First published in German as "Kasuistischer medizinischer Unterricht im 16. Jahrhundert. Die paduanischen *collegia*", in: *Innsbrucker Historische Studien* 39 (2024), pp. 41–54. Translated into American English by Logan Kennedy and Leonhard Unglaub.

[1] Historical surveys in Nancy Siraisi, "The faculty of medicine", in: *A history of the university in Europe*. Vol. 1: *Universities in the Middle Ages*, ed. by Hilde de Ridder-Symoens, Cambridge 1992, pp. 360–387; eadem, *Medicine and the Italian universities 1250–1600*, Leiden, Boston, and Cologne 2001; Laurence W. B. Brockliss and Colin Jones, *The medical world of early modern France*, Oxford 1997, pp. 85–107; Nancy G. Siraisi (ed.), *Medicine and the Italian universities 1250–1600*, Leiden 2001.

M. Stolberg, *Medicine and the Body in Early Modern Europe*, Palgrave Studies in Medieval and Early Modern Medicine,
https://doi.org/10.1007/978-3-031-84556-7_9

healers who were responsible for the lion's share of medical care in both urban and rural areas. In addition, as this development unfolded, greater numbers of municipal authorities began to hire doctors of medicine at a fixed salary as town physicians, thus ensuring a constant availability of local, learned medical care.[2] The growing demand for professional medical services along with the status of the municipal physician and the accompanying guarantee of a basic income made becoming a physician an attractive career option in the sixteenth and seventeenth centuries for young men from the upwardly mobile urban middle classes. Thanks only to their medical training, which they often completed far from home without the benefit of wealthy parents or the support of a network of relatives, they were able to establish a flourishing practice that supported them, or even allowed them—if things went very well—to amass great wealth.

Among physicians, there was widespread agreement that the prerequisite for a successful medical practice was sound diagnostic, prognostic, and therapeutic knowledge and skill, which would guarantee therapeutic outcomes superior to those produced by the many competitors. While much knowledge germane to practice could be gained from authorities such as Hippocrates, Galen, or Avicenna, who continued to play an important role in medical studies, this book knowledge was now increasingly perceived as insufficient. It was the growing conviction that one best learned to diagnose and successfully treat the pathological changes in the body from following adept practitioners and by gaining experience at the bedside. In the sixteenth and early seventeenth centuries, many aspiring physicians made the move to northern Italian universities not only because of the outstanding anatomical training that was offered there, but also because practical teaching was much more highly developed and prioritized there compared with the universities north of the Alps.[3]

[2] Cf. Martin Kintzinger, "Status medicorum. Mediziner in der städtischen Gesellschaft des 14. bis 16. Jahrhunderts", in: *Städtisches Gesundheits- und Fürsorgewesen vor 1800*, ed. by Peter Johanek, Cologne 2000, pp. 63–92; Ruth Schilling, Sabine Schlegelmilch, and Susan Splinter, "Stadtarzt oder Arzt in der Stadt? Drei Ärzte der Frühen Neuzeit und ihr Verständnis des städtischen Amtes", in: *Medizinhistorisches Journal* 46 (2011), pp. 99–133; Michael Stolberg, *Gelehrte Medizin und ärztlicher Alltag in der Renaissance*, Munich 2021.

[3] On medical education in Padua see Giuseppe Ongaro, "La medicina nello Studio di Padova e nel Veneto", in: *Storia della cultura veneta. Dal primo Quattrocento al Consilio*

Bedside Teaching

Medical historians have largely focused their attention on the emergence of bedside teaching. Giovanni Battista da Monte has been considered the founder of bedside teaching. He was already teaching his students at the beds of patients in the Ospedale San Francesco in Padua around 1540.[4] The notes taken by his students not only mention the numerous visits to patients in San Francesco and quote statements Da Monte made at these occasions. They also show that Da Monte regularly had them join him on his rounds so that they could study the course taken by the disease when the patient was under medical treatment. So we learn of students visiting a consumptive patient with empyema at least ten times, a boy with scabies sixteen times, and in connection with a clergyman who was suffering as a result of mercury treatment for his "French disease", they even recorded a seventeenth visit at the hospital.[5]

The notes taken by students furthermore attest to the fact that their professor did more than explain his diagnosis and treatment. They could also acquire and practice diagnostic skills. They examined the urine of dozens of patients—uroscopy was the most important diagnostic procedure at the time—and they could compare their conclusions with those of the professor. They tried their hand at the difficult art of pulse diagnosis: here it was necessary to judge not only frequency and possible arrhythmia, but also the strength and speed of each pulse wave as it surged. And they

di Trento, vol. III/III, Vicenza 1981, pp. 75–134; Cynthia Klestinec, "Medical education in Padua. Students, faculty and facilities", in: *Centres of medical excellence? Medical travel and education in Europe, 1500–1789*, ed. by Ole Peter Grell, Andrew Cunningham, and Jon Arrizabalaga, Abingdon and New York 2010, pp. 193–220; Michael Stolberg, "Training future practitioners: Medical education in sixteenth- and early-seventeenth-century Padua and Montpellier from the students' perspective", in: *Medical education. Historical case studies of teaching, learning, and belonging in medicine in honour of Jacalyn Duffin*, ed. by Delia Gavrus and Susan Lamb, Montreal and Kingston 2022, pp. 112–135.

[4] Andrea Comparetti, *Saggio dela Scuola Clinica nello Spedale di Padova*, Padua 1793; Giuseppe Ongaro, "L'insegnamento clinico di Giovan Battista da Monte (1489–1551): una revisione critica", in: *Physis* 31 (1994), pp. 357–369; Michael Stolberg, "Bedside teaching and the acquisition of practical skills in mid-sixteenth-century Padua", in: *Journal of the history of medicine and allied sciences* 69 (2014), pp. 633–661.

[5] Giovanni Battista Da Monte, *Consultationum medicarum opus absolutissimum*. Ed. by Johannes Crato, Basel 1565, coll. 455–461: "De empyico et phthisico in hospitali, visitationes decem", col. 885, "in hospitali, in XVI. accessione" and coll. 867–869, "Haec cura facta fuit [...] in decima septima visitatione hospitalis, quae dedit occasionem octo lectionibus de morbo gallico."

learned to palpate the abdominal area of the patient with their own hands. For generations, medical historians claimed that this diagnostic procedure, which is very common today, was not practiced in the sixteenth and seventeenth centuries, yet the student notes of the time show that it was common.[6]

There are also indications of practice-oriented teaching at the bedside at other Italian universities at the time. The *Processus,curationes et consilia* of Elideo Padoani (died 1575) were avowedly based on what he communicated to the students who followed him in his practice.[7] From Ferrara we have student notes about the cases of dozens of patients the students saw, probably in the 1540s, when they followed Antonio Musa Brasavola (1500–1555) and other professors sometimes daily or even twice in a single day.[8]

The *Collegia*

There is, however, another innovative form of practical, casuistic teaching for which the Upper Italian universities, especially Padua, were likewise famous and which historians have so far almost completely overlooked: the so-called *collegia*. In a *collegium* several professors came together to discuss a specific case in front of students. At the outset, one of the professors would present the patient's history either in his own words or as a written report which he read aloud or summarized. Then each professor in attendance would submit his own medical opinion verbally, one after the other. From the complaints that had been described and, if applicable, from the externally visible signs of disease, the professors would draw conclusions about the pathological changes inside the body, and from this they would determine which remedies were most suited to combatting the causes of the disease and healing the patient.

[6] Detailed analysis in Stolberg, Bedside teaching [see note 4].

[7] Elideo Padoani, *Processus,curationes et consilia in curandis in particularibus morbis quae prosperos habuerunt eventus [...] medicinae candidatis in praxis eum sequentibus communicata*. Ed. by Johannes Wittich, Leipzig 1607.

[8] Biblioteca Ariostea, Ferrara, Collezione Antonelli, ms. 531; this manuscript was already mentioned by Cesare Menini, "'Curationes A. M. Brasavoli'. Contributo alla conoscenza delle opere di Antonio Musa Brasavolo come medico pratico", in: *Rivista della storia delle scienze mediche e naturali* 43 (1952), pp. 255–261.

In this way, the *collegia* fulfilled an eminently important didactic function. As they did in their oral explanations during clinical rounds, the professors communicated central and fundamental methodological skills. And they did so drawing on concrete cases characterized by clinical pictures that the students were likely to encounter in a similar form later on in their own practice. The professors regularly gave detailed reasons for their assessment, went through the various possible causes as one would when performing a differential diagnosis, and carefully drew their conclusions about the pathological changes inside the body on the basis of the external manifestation of the disease, the patient's temperament and lifestyle, the medical history, the symptoms, etc. Paduan medicine became famous for this clinical method, which was also considered the key to a successful practice down the line. After all, learned medicine as a whole was in agreement that if treatment was to allow the patient to recover and remain healthy on a long-lasting basis, it was not to remain limited to a mere "palliative" treatment as it was called then, that is a treatment in a literal sense only "cloaked" the disease.[9] Rather, the aim was to identify the causes of the complaints inside the body and to eradicate them, that is rid the patient of them through a "radical" treatment that got to the root causes.[10] And so, in a dedicatory letter to the Danish physician Severinus (1542–1602), we find the printer Konrad Waldkirch (1549–ca. 1615) emphatically praising the practice of physicians discussing their conclusions about the nature, causes, and treatment of the sick in front of students, a practice that was particularly well-established in Padua. These joint consultations taught students how to think about many different things in many different ways, and gave the patients hope of a complete recovery.[11]

[9] As Georg Handsch, who studied in Padua in the 1550s, wrote, "palliare" meant to "cover up the disease for some time" ("die Krankheit auff ein Zeit vertuschen"), ÖNV, Cod. 11206, fol. 135v.

[10] See Stolberg, Gelehrte Medizin [see note 2], pp. 53–54.

[11] Vettore Trincavella, *Consilia medica*, Basel 1587, dedicatory letter by the printer Konrad Waldkirch to Petrus Severinus, Basel, April 1, 1577: "Haeret enim adhuc animo tuo memoria celebrati moris illius et frequentati praecipue in schola patauina, quem doctores vsurpant in conferendis iudiciis de morborum natura, caussis et remediis, circa aegrorum lectulos. Quae consuetudo consultandi praeterquam quod discentibus de multis et variis rebus consideratione dignis diuerso modo cogitandi occasionem subministrat, non minus etiam atque adeo multo magis, aegris magnam spes regressus ad pristinum et inculpatum naturae statum facit." Severinus had studied medicine in Padua in the 1560s.

The *collegia* took place frequently. In a manuscript that is held today in Wolfenbüttel, an unknown German student recorded statements made by professors in almost fifty *collegia*, which he attended in Padua in the 1550s over the course of two years.[12] According to student notes, the number of participating professors varied. Often, there were only three—"tres faciunt collegium" as one still hears today—but sometimes there were four or five. In 1556, for example, when the Baron of Dietrichstein was suffering from vertigo and other complaints, Gabrielle Falloppia, Alvise Bellocati, Antonio Fracanzano, Francesco Frigimelica, and Vettore Trincavella gave their opinion one after the other.[13] Occasionally, physicians and surgeons who did not have a teaching position would weigh in as well.

The place of their gathering—judging by the student notes—was usually the house of one of the participating professors. In the 1550s this was principally the house of Francesco Frigimelica (1490–1558), but occasionally it could also be that of Trincavella or Oddo degli Oddi (1478–1558). Then again, the Wolfenbüttel manuscript mentions three *collegia* that were held outside in the sun, and others at further locations, and once professors and students even got together at an *osteria*.[14] A *collegium* could be held based on an external, written report, but even when it was based on a case at hand, the notes do not usually tell us whether those seeking medical advice were present at the *collegium*. The place where the *collegia* were held suggests that at least in the case of bedridden patients, students and professors first went to the patient's bedside and then got together for discussion outside or back at the professor's house. Here and there we do find indications of patients, however, who were introduced in person at a *collegium*, for example when a professor began his explanations with the words, "The man you see before you is in his forty-fifth year".[15]

The *collegia* share common characteristics with the then widespread and familiar practice of joint medical consultations, from which they may have developed. Patients from the nobility quite often demanded that

[12] Herzog August Bibliothek, Wolfenbüttel, Cod. Guelf. 20. 22. Aug. 4to. The total number of *collegia* results only from the table of contents. Notes on the cases n° 10 to 20 are no longer extant.

[13] Ibid., foll. 46v-49v, March 3, 1556.

[14] Ibid., foll. 26v-29v, foll. 53–56r, foll. 61v-65v, and foll. 127r-130v.

[15] Wellcome Collections, London, Western Manuscripts, Ms. 602, no page numbers, the name of the professor is not mentioned.

several physicians come to their bedside and give their opinion. Others would send a lengthy written account of their medical condition to a faculty or a medical college, requesting collective medical advice by letter.[16] The medical opinions offered at the *collegia* by the participating professors were later sometimes also published in printed collections of *consultationes* and *consilia*. The term *consultatio* was used metonymically for the written-down conclusions drawn from such mutual deliberations, and *consilium* was the umbrella term for all medical advice delivered in writing, regardless of whether it came from a single physician, as was usually the case, or from a group of physicians or a medical panel. Thus, in the collections of Vettore Trincavella's *consilia* we find notes about a whole series of *collegia* carried out by Paduan professors, alongside numerous consultation letters which Trincavella (1496–1568) had sent to individual patients as part of his epistolary practice.[17] This has led to to misunderstandings in historical research, when notes about *collegia* were taken to be common *consilia*.[18] However, just looking at the form they take, the notes about *collegia* differ quite substantially from the surviving written *consilia* and *consultationes* in the traditional sense. The notes were taken by students who were in attendance, usually clearly identifying the statements the participating professors made one after another regarding the case at hand. The oral character also shines through when students reproduced verbatim, in the first person, what a professor had said. It is further striking that the professors' names are often spelled in a very unconventional manner, suggesting that the students had only heard and not read them. "Bellocati", for example, is easily rendered "Pelocato",

[16] See, for example, Universitätsbibliothek Erlangen, Ms. B 323/1, foll. 56r-60r, *consilium* by the Nürnberg *collegium medicum* for the eye disease of Heinrich Mülegg (detailed summary in www.aerztebriefe.de/id/00026576); the open-access database of the longterm project "Early modern physicians' letters" in Würzburg offers various other examples (www.aerztebriefe.de).

[17] Trincavella, Consilia [see note 11].

[18] Drawing on student notes on Paduan *collegia*, which were published together with *consultationes* and *consilia* by Giovanni Battista Da Monte and Girolamo Capivaccia, Monica Calabritto has attempted to show a profound change in the genre of *consilium*. In doing so, she has ignored the fundamental difference, however, between a written *consilium* and student notes on the oral contributions to a joint case discussion; cf. Monica Calabritto, "Curing melancholia in sixteenth-century medical *consilia* between theory and practice", in: *Medicina nei secoli* 24 (2012), pp. 627–664.

"Falloppia" or, Latinized, "Fallopius", becomes "Phallopius", whereas "Frigimelica", also called "Frisimelica", is turned into "Phrysimelica".[19]

An Exemplary *Collegium*

In the following, I will illustrate how a *collegium* was carried out and what its outcomes were using the specific example of one that took place in Padua on 20 March 1559 regarding a seventy-year-old man with urinary complaints, as documented by an unknown student. It was a comparatively straightforward case about which the participating professors were largely in agreement. For precisely this reason, the notes show all the more clearly the systematic, methodological approach and the nuances of the diagnostic, prognostic, and therapeutic approach for which Padua was famous and which was taught in the *collegia* so that students could apply it in their own future practice.[20]

At the start of the *collegium* in question, the anatomist and surgeon Gabrielle Falloppia presented the case with the help of a written description—at the margin we see "historia". The patient had been suffering for two years from a frequent urge to urinate and from pain during urination. He had already of a burning sensation during urination before, but then it was only temporary and the pain was not intense. The urine was sometimes light-colored and sometimes yellow, and it contained visible, bran-like ("furfum") and sandy admixtures and some matter that was similar to ground eggshells. Sometimes the urine had the consistency of semen, but it did not stink. The patient's natural "habitus" was hot and dry, but due to his age, his heat was diminished ("remissa"). He was slender and of a hot and dry temperament. His brain was also characterized by this temperament, as could his light sleep and his intelligence indicated, though it was likely on the border between dry and moist.

[19] Herzog August Bibliothek Wolfenbüttel, Cod. Guelf. 20. 22. Aug. 4to, fol. 44r, fol. 24v, and fol. 46v.

[20] Biblioteca comunale di Siena, Ms. C IX 32, foll. 1r-5r; the author is unknown but it certainly was not Giulio Mancini, to whom large parts of this manuscript must be attributed. He studied in Padua later only and probably he copied the notes on the first three *collegia* that are documented here from the notes of a student who did witness them.

Next, Falloppia turned to the causes, to the "causarum indagatio", as is indicated in the margin. He explained that the burning during urination could come from changes in the urethra itself or from the urine. For its part, the urine could be of an acrid, burning constitution or it could contain admixtures. Because the patient excreted sand, Falloppia considered the decisive cause to be acrid, burning admixtures. However, in his assessment, changes in the urethra itself could also be a contributing factor. There were no signs of a fully formed urethral ulcer, such as a discharge of pus and visible changes in the penis. However, he suspected that the urethra was worn and its lining had become thin and that the glands that covered it with an oily substance were weak, which was not so surprising in a man of his age with a hot and dry temperament. The patient's hemorrhoidal bleeding further contributed to a drying out of these parts as well as of the entire body.

Falloppia then went on to discuss the nature and source of the acrid admixtures in the urine. The patient was very active, had a weak stomach and concocted his food poorly. The food, having been insufficiently concocted and thus remaining raw, caused diarrhea and produced a slimy substance—instead of healthy chyme, as is implied. This slimy substance made its way to the liver, where it was heated and burned, just like the other humors. This produced coarse bile, which became visible in the urine as sand. The solid, eggshell-like particles came from a solidification in the form of small concretions in the bladder or urethra. For Falloppia, this also raised the question about whether the man had bladder stones. However, Falloppia did not see any indication of this: no sensations of heaviness, no stabbing pain, no urine that remained in the bladder after urination.

Now Falloppia turned to treatment—the word "curatio" is written in the margin. The aim had to be to remove the morbid matter ("materiam peccantem") and to strengthen the affected parts. This was difficult in a patient of this age, but one would have reason to hope for improvement if the influx of the mucilaginous and bilious burnt matter from the liver could be stopped, if the hot and dry *intemperies* of the liver could be corrected, and the stomach could be strengthened. Going by his analysis of the pathological processes in the body and the *indicationes* he derived from them, Falloppia arrived at extensive therapeutic advice, which can only be briefly summarized here. He believed bloodletting was unnecessary because of the hemorrhoidal bleeding. He thought an array of laxative medications such as cassia, rhubarb, and senna would serve to expel the morbid matter. The stomach was to be strengthened with warming remedies. Further favorable results could come from using hot springs like those at Montecatini or at Villa near Lucca. Of thousands of people who had used them, he had not heard of two who had not been cured of their stomach weakness there. Moreover, emetics were worth a try. And a local treatment of the painful penis with oily, moistening remedies was recommended, both externally, and internally, by injecting remedies such as butter or almond oil.

After Falloppio, Alvise Bellocati (alias Aloysius Bellocatus, 1501–1575) spoke. He explicitly agreed with Falloppia's explanations.[21] He too ascribed the complaints to acrid admixtures in the urine and saw no indication of stone disease. Rather, the substance was heated in the kidneys or bladder and then solidified. As the pathogenetic *causae efficientes*, he identified an *intemperies* of the stomach, liver, and kidneys. The prognosis was not favorable because according to Hippocrates's *Aphorism* 6.6, complaints of the kidneys and bladder were difficult to heal in very old people. As much as possible, treatment had to counter the generation of morbid matter and to weaken its effects on the urethra. He recommended evacuating the morbid matter through other pathways, by giving laxatives and stimulating hemorrhoidal bleeding, as well as addressing the *intemperies* of the organs. The difficulty was that correcting the hot and dry *intemperies* of the liver and kidneys with cooling remedies threatened to intensify the cold *intemperies* of the stomach and vice versa. However, as Averroes had emphasized in the fourth book of his *Colliget*, wise Nature, in her efforts to preserve the body, knew how to separate out those parts that were useful to a certain organ and to lead them directly to that organ. Specifically, Bellocati recommended absinth, which had a gentle, warming effect and fortified the body noticeably, as his listeners knew from their reading of Galen's works. Contradicting Falloppia, he expressed a certain skepticism regarding the drinking of healing waters from a hot spring. They did not always have beneficial effects. But he did not want to advise against the waters of Villa near Lucca altogether. When one administered them with manna or sugar, they could help evacuate the acrid matter through the bowels and they were also good for the liver as long as the stomach had been sufficiently strengthened first. Additionally, their external application could help warm the stomach and cool the

[21] Ibid., fol. 4v: "Haec est sententia mea confirma[n]da ea, quae ab Excell.o Viro proposita sunt, potius n[empe?] ei subscribo, quam aliud addere velim etc."

liver and kidneys. The stomach could be strengthened, he recommended, with sweet confections and such. Pepper, on the other hand, was too hot and would make the urine even more acrid. Moreover he recommended a local treatment of the penis and urethra, among others using such moistening substances as butter and almond oil. The patient was to continue with a moderate diet and to drink thin, not too strong wine, though it should not be too watered down either.

Antonio Fracanzano (1506–1567) spoke last. He largely concurred with the opinions of the previous physicians. The acrid urine that was abrading the urethra came from the excessive heat of the liver and kidneys and from a cold stomach in his view. The cold stomach insufficiently concocted the food and the overheated liver resulted, among other things, in the excretion of burnt blood through the hemorrhoids. Furthermore, the excretion of salty matter in the urine showed that the rest of the body was filled with this matter. Such salty humors were characteristic for old men and could not be evacuated entirely. Thus, a complete cure was impossible and he hoped that the patient would accept strong remedies as weak remedies were ineffective against humors that were spoiled to this extent. Fracanzano recommended remedies like cassia and hiera picra, to open the hemorrhoids and evacuate the matter in this and other ways. The stomach could be warmed and strengthened by drinking whey on a daily basis. He considered the water of the hot spring at Villa near Lucca to be indicated. If the patient wished, he could also take emetics twice a month. Suitable nutrition and dry wine could do their part. In the fall, he should take purgatives again. He would not, however, regain the health he once enjoyed.

As mentioned at the outset, the case was a relatively simple one with only one central symptom, pain during urination. In other cases, the clinical picture was far less self-evident. For example, in a case where a middle-aged man complained of headache, noises in his ears, vertigo, burning catarrhal discharge, and a loss of appetite, finding a cause of the illness that covered all symptoms and was in line with the prevalent medical ideas of the time was not as simple a matter.[22] The notes about the comparatively straightforward case of the elderly patient experiencing a burning sensation during urination thus give an all the more vivid impression of the systematic, methodological approach that was being

[22] Herzog Augustbibliothek Wolfenbüttel, Cod. Guelf. 20. 22. Aug. 4to, foll. 81v-85v.

demonstrated to the participating students by professors with the help of this representative case. This was an approach the students would later have to adopt for their own practice if they wanted to be successful. At the center was always an analysis of the presumed causes of illness, more exactly, the pathological—today we would say pathogenetic—processes and changes inside the body. This analysis was based on a precise observation of the symptoms and the externally visible changes, a determination of the predisposing and triggering factors, for example temperament and way of life, and it often involved a careful differential diagnosis, that is a ruling out of possible alternative causes and explanations. With this grounding in a precise assessment of the nature of the disease and the pathological processes occurring inside the body, the participating professors then recommended various therapeutic measures and medications which were to counter the pathological changes and processes and thus allow for a causal treatment.

The professors' diagnostic and therapeutic assessments shown in this representative case reflect the predominant Galenic understanding of disease.[23] While the opinion obstinately persists even in professional research in the history of medicine that physicians back then related most diseases to an imbalance of the four humors and their qualities in the body, student notes pertaining to the *collegia* along with the hundreds of case histories published at the time make it clear that this is a myth. One has to search as if for the proverbial needle in the haystack for cases in which a physician actually deemed a disease to have been caused by a disproportion in the humors or qualities in the body, for example an excess of black or yellow bile. Rather, the vast majority of diseases was ascribed to more or less specific morbid matter that was disseminated in the blood and/or had settled in certain places in the body.

Successful treatment therefore demanded that the nature of this morbid matter be identified and specifically targeted, and that it be evacuated from the body or at least its course in the body diverted so that it no longer flowed to the affected body part that was already weakened.

In order to bring about a permanent cure, the physician also had to identify and combat the source of ongoing production and/or accumulation of morbid matter in the body, however. During the process of diagnosis at least one of three fundamental pathogenetic processes was

[23] On the following, see my detailed account in Stolberg, Gelehrte Medizin [see note 2], pp. 125–337.

regularly taken to be at play and often physicians assumed that two or all three interacted and reinforced each other. Physicians very often found that the stomach and its concoctive heat were too weak and thus insufficiently concocted the raw food that was consumed each and every day. The blood that was produced in the liver from the insufficiently concocted chyme and the resulting humors were, in turn, of an inferior, morbid nature and could continue to spoil inside the body and become even more malignant. A second cause that was frequently diagnosed was an *intemperies*, an imbalance of the qualities in a specific, important organ—*nota bene*: not in the body as a whole—which disturbed the function of this organ and/or changed the humors that flowed through it in a pathological manner. One of the most frequent diagnoses was, as in the case I have described in detail, that of an overheated liver which burned the chyme and the blood that was produced from it as well as the other humors, thus lending them an excessive acridity. This helped explain not only pain during urination but also the formation of cancerous ulcers and other severe changes in the skin for example, or scurvy, in which the morbid matter "ate through" the gums.[24] Stringy, tough humors that had thickened due to the evaporation of liquid or because of admixtures of raw mucus played a major role in the third frequently postulated cause of disease, a so-called "obstruction" in the body. Obstructions hindered the free movement of the humors but above all they interfered with natural excretion, thus leading to an accumulation of raw, impure substances in general and, in the event of illness, of a morbid matter in particular. Of all causes of disease in women aged fifteen to sixty, an "obstruction" of the "monthly purification" was considered the most prevalent and powerful. Further important and recognized explanatory models came into play in different ways in the *collegia*, depending on the case, for example that of decreasing *calor innatus* or vital heat and the idea of vapors and trails of smoke rising up from the abdominal area and impairing brain function.

[24] See, for example, Wellcome Collections, London, Western Manuscripts, Ms 567, foll. 18r-20v, "Collegium pro muliere patiente defedationem faciei, qui affectus impetigo dicitur"; the unidentified professor or physician who presented the case, assumed that the liver was very hot and produced burnt fluids which in turn affected the facial skin.

An Academic Ritual

More than anything, the *collegia* were an original and highly valued didactic instrument. At the same time, seen from the perspective of social and cultural history, they had the quality of an academic ritual. They very clearly expressed the hierarchical relationships between the professors in front of a group of students. After all, the order in which the professors were given the floor was in no way random. Without fail, the *collegium* was opened by the professor on the bottom rung. He was the one who had to introduce the case and be the first to give his opinion. His contribution was then followed by the others, in turn, and the highest-ranking professor had the last word, both literally and figuratively.

This sequence corresponded with the common practice of joint consultations at the bedside of noble patients. Here, the first to speak was the youngest and least experienced and then came those with more experience. The *collegia* differed in one essential aspect, however: the order of priority among the professors did not reflect their practical experience or any special competence regarding the medical condition in question, but exclusively the status of their professorship. For a long time, the professorship for theoretical medicine ranked highest and was most highly paid. In the course of the sixteenth century in Padua and at a few other universities, professors of practical medicine came to be remunerated equally or were even awarded higher pay. Far below on the ladder were professors for *materia medica*, that is, those who taught about medicinal plants, and those who taught anatomy and surgery. In Padua, their salaries were only a fraction of those awarded to the higher-ranking professors. Although he taught anatomy and surgery as well as *materia medica*, Gabrielle Falloppia received only 200 gulden per year in the 1550s, whereas Vettore Trincavella, as the university's first professor of practical medicine, received 1200 gulden.[25]

What this meant for the holders of the less respected chairs, quite apart from the financial aspects, can be illustrated by the example of Gabrielle Falloppia. In the middle of the sixteenth century, Falloppia was a European celebrity. One of the best anatomists of his time, he had not only, like his predecessor Andreas Vesal, brought to light many errors of traditional Galenic anatomy, but had, unlike Vesal, made a whole series of

[25] Michael Stolberg, *Gabrielle Falloppia, 1522/23–1562. The life and work of a Renaissance anatomist*, London 2023, p. 147.

anatomical discoveries. The most well-known of these is his precise depiction—the first of its kind—of the ovaries and fallopian tubes—which still today carry his name. Moreover, he was in high demand as a surgeon and not least of all, he was a highly learned humanist who was able to read the Greek authors in the original. Students from all over Europe sought him out so that they might participate in his anatomical demonstrations. His medical advice was sought by princes and the pope himself.[26]

And yet, during the *collegia* he was compelled to speak first, as was his successor Girolamo Fabrizi d'Acquapendente later on, who was hardly less famous. The only exception was if an even lower-ranking colleague from outside the university was taking part. As described above, Falloppia had to present the case and then submit his medical opinion in front of everyone else like a beginner, like the least-experienced physician at the bedside of an aristocratic patient. The other professors took their turns after him, which meant that he always ran the risk that they would—before the assembled students—declare his judgment to be incorrect, or at least part of it. In the surviving student notes that I have examined, I have not been able to find any major, fundamental controversies regarding the nature of a disease or its causes. The professors who spoke after Falloppia did object to some of his recommendations for treatment, however. They might discourage bloodletting when he had advised it, the might give preference to different medications, or—as in the case we have seen—might question the positive effects of thermal spring water, which Falloppia, going by his own experience it seems, praised as beneficial. Even with surgical cases, in which he had far more experience and expertise than his colleagues, he could not defend himself against their divergent opinions—he was not granted another chance to speak.

All this unfolded in a highly toxic atmosphere—as we would say today—because Falloppia's fellow professors begrudged him his popularity and his success. Invoking the statutes, they were even able to see to it in those years that Falloppia was no longer allowed to combine—as Andreas Vesal had done before him—the manual work of dissecting a human body with explanatory demonstrations and a preceding general anatomical lecture. This lecture was claimed by the aging Trincavella, who was becoming a laughing stock among students because of his declining

[26] See Stolberg, Falloppia [see note 25], pp. 147–154, for a detailed account.

memory.[27] All that fell to Falloppia now was the manual task of dissecting. In the end, the episode turned into a fiasco. The students raised such a ruckus that Trincavella could no longer make himself heard. They kept chanting, "We want Falloppia". The lecture and the subsequent demonstration led by Trincavella had to be cut short. It does not take great imagination to conceive of how humiliating the incident would have been for Trincavella and how it must have overshadowed the relationships between the participating professors. And we can also gain a sense of the intensity of the animosities between the professors by considering the events that took place after Falloppia's early death. Just a few days after Falloppia died, one of his greatest adversaries in Padua, Bassiano Landi, professor of theoretical medicine, was attacked by unknown assailants and soon succumbed to his severe injuries. It was said that a letter had fallen into the wrong hands, in which Landi unabashedly expressed his joy over Falloppia's death. The murder was never solved. Contemporaries assumed that Falloppia's students killed Landi.[28]

Conclusion

To summarize, it can be said that the Paduan *collegia* served and combined multiple functions in an original and innovative manner. Just like the joint consultations at the bedside of high-status patients, convening well-known, experienced physicians promised the best possible insights into the nature of a disease and the pathological processes inside the body that were at its root and, based on this, gave hope of the most effective treatment. At the same time, *collegia* were an academic ritual in which the hierarchical relationships between participating professors and their respective fields of study and practice were continually renegotiated, with personal animosities sometimes also playing a role. More than anything, however, they were an outstanding didactic instrument, and this is what brought them their fame. With good reason, contemporaries

[27] Universitätsbibliothek Erlangen, Trew, Purkircher, Nr. 4, letter from Georg Purkircher to Joachim Camerarius; the letter carries the date February 18, 1563 but by all appearances Purkircher made a mistake and the letter was written on March 18, 1563 only; see www.aerztebriefe.de/id/00002573 for a detailed summary of the letter by Michael Clement and Tilmann Walter.

[28] Stolberg, Gabrielle Falloppia [see note 25], pp. 147–154 and pp. 162–163.

recommended that the practice of professors discussing cases together in front of students be introduced north of the Alps as well.[29]

Even with a very broad definition of the concept, we would gain a very limited understanding of the *collegia* if we casted the students' experience of them as a "rite of passage". They were passive members of the audience, unlike during clinical teaching at the bedside and during anatomical dissections, in which they could sometimes play an active part. Along with the joint visits at the bedside, the *collegia* without a doubt constituted an important stage in a student's medical training and socialization, however, one which we often find them recalling in retrospect. The professors' extensive explanations of each case of illness taught students not only elementary diagnostic and therapeutic skills. They were also, as witnesses, being schooled in the argumentative exchange by which divergent medical opinions were negotiated. This helped foster a skill that would be indispensable in the future if they wanted to be able to hold their own in joint consultations about the diseases of high-ranking (and accordingly high-paying) patients, who might even be present in the room. And not least of all, from the repeated experience gained in the *collegia*, aspiring physicians learned that even eminent professors with extensive practical experience could arrive at different conclusions when examining cases in detail. On the one hand, this reflected the uncertainty of medical knowledge, and on the other, it illuminated the leeway one had in devising one's own treatment. In this respect, the *collegia* helped students cultivate an ability that in the late eighteenth and nineteenth centuries would be aptly described by physicians as medical *savoir faire*.[30]

[29] Trincavella, Consilia [see note 11], dedicatory letter by Konrad Waldkirch to Petrus Severinus, Basel, April 1, 1577.

[30] Samuel Gottlieb Vogel, "Einige allgemeine Bemerkungen über das Sçavoir faire in der medicinischen Praxis", in: *Journal der practischen Arzneykunde und Wundarzneykunst* 1, 3. Stück (1796), pp. 265–324; on the genre, see also Iris Ritzmann, "Der Verhaltenskodex des 'Savoir faire' als Deckmantel ärztlicher Hilfslosigkeit? Ein Beitrag zur Arzt-Patient-Beziehung im 18. Jahrhundert", in: *Gesnerus* 56 (1999), pp. 197–219.

The Concept of Innate Heat in Avicenna's *Canon Medicinae*

Introduction

After Gerard of Cremona had translated it into Latin, Avicenna's *Canon medicinae* quickly became a standard medical textbook at European universities and would remain so for centuries. Taking natural-philosophical questions into account, it comprehensively and systematically treated all subdomains of medicine and linked them together to form a homogenous whole. The *Canon* soon took on a key role in medical training and practice. Dedicated to physiology and the structure of the human body, the first book in particular, along with a proliferating literature of commentary, was a fundamental and central work in the medical training curriculum at universities. Thus, it can be said that the Latin

First published in German as "Die Lehre vom 'calor innatus' im lateinischen *Canon medicinae* des Avicenna", in: *Sudhoffs Archiv* 77 (1993), pp. 33–53. Translated into American English by Logan Kennedy and Leonhard Unglaub.

M. Stolberg, *Medicine and the Body in Early Modern Europe*, Palgrave Studies in Medieval and Early Modern Medicine,
https://doi.org/10.1007/978-3-031-84556-7_10

version of the *Canon* came to assume its own, largely independent significance in the history of Western medicine. As a source *sui generis*, this version will serve as the basis for the following considerations.[1]

Considering that Avicenna's significance in European medicine is undisputed, it is surprising that there has so far been only insubstantial study of his medical writings, except for some studies of particular, closely circumscribed areas.[2] A comprehensive study of the fundaments of Avicenna's physiology and pathology has yet to be carried out.[3] The main reason for this may be that Avicenna's principal achievement has been seen in his synthesis and systematization of Galenic and Aristotelean

[1] On the history of the translation and reception of the *Canon* see Nancy G. Siraisi, *Avicenna in Renaissance Italy. The Canon and medical teaching*, Princeton 1987; on its early reception in Italian medicine see Konrad Goehl, "Guido d'Arezzo als Avicenna-Leser", in: *Würzburger medizinhistorische Mitteilungen* 1 (1983), pp. 23–36, and idem, *Guido d'Arezzo der Jüngere und sein "Liber mitis"*, Pattensen/Han. 1984, pp. 23–27. I cannot address the problems of translation into Latin here; see, however, Hakim Abdul Hameed, "Gerard's Latin translation of Ibn Sina's Al-Qanun", in: *Studies in Islam* 8 (1971), pp. 1–7. I will quote the *Canon* from the edition which Benedetto Rinio made on the basis of Andrea Alpago's improvements on Gerhard of Cremona's translation and which came out in Basel, in 1556 (*Avicennae medicorum Arabum principis, Liber canonis, De medicinis cordialibus, et Cantica*, Basel 1556 (repr. Teheran 1976); compared to the 1507-edition (*Liber canonis Auicenne reuisus et ab omni errore mendaque purgatus summaque cum diligentia impressus*, Venice 1507 (repr. Hildesheim 1964)) it is, among others, much easier to read.

[2] For example, Antje Bornemann, *Das Krankheitsbild der Phrenitis in der Medizin des arabischen Mittelalters unter besonderer Berücksichtigung von Ibn Sina (980–1037)*. Diss. med. Bonn 1988; Julius Hirschberg, *Die Augenheilkunde des Ibn Sina aus dem Arabischen übersetzt und erläutert*, Leipzig 1902; Erhart Kahle, *Avicenna (Ibn Sina) über Kinderkrankheiten im Kinderregimen seines Qanun*, Erlangen 1980; Gernot Rath, *Die Anatomie des Avicenna und die Nomina anatomica in der Canonübersetzung des Gerhard von Cremona*. Diss. med. Bonn 1948.

[3] Only rough overviews without a thorough analysis of the basic concepts are offered by Albert Nader, "Avicenne médecin", in: *Orientalische Kultur und europaisches Mittelalter*, ed. by Albert Zimmermann et alii, Berlin 1985, pp. 327–343, and by Heinrich Schipperges, *Eine "Summa Medicinae" bei Avicenna. Zur Krankheitslehre und Heilkunde des Ibn Sina (980–1037)*, Berlin 1987; two modern translators approach these issues somewhat more closely in their commentaries on the first book of the *Canon*, namely Mazhar H. Shah, *The general principles of Avicenna's "Canon medicinae"*, Karachi 1966, and O. Cameron Gruner, *A treatise on the Canon of Medicine of Avicenna, incorporating a translation of the first book*, London 1930; their translation and commentary is strongly shaped by their desire to highlight the importance of the teachings of the *Canon* for modern medicine and philosophy, however, leading them to one-sided and anachronistic interpretations.

concepts which in themselves were well-known.[4] Yet, for university physicians who were increasingly interested in such questions, the fundaments of physiology, pathology, and pharmaceutics as presented in the *Canon* constituted a central reference point for their thinking, one with which they engaged productively as they explored medical topics.[5] Already with Galen, the fundaments of medicine are in no way readily accessible and logically consistent. The *Canon* furthermore connects the Galenic teachings with a specific interpretation of Aristotelean natural philosophy as had become pervasive among Islamic scholars around the turn of the millennium. The most characteristic aspect here—next to the influence of the Islamic doctrine of creation—is the integration of neoplatonist influences in the interpretation of Aristotle.[6] Thus, to understand the history of the reception of the *Canon*, it is essential to first conduct an independent analysis of the ideas he formulated in it regarding fundamental life processes.[7] In the following, I will consider these ideas more closely, looking at the theory of the *calor innatus* and the closely related concepts of *spiritus innatus* and the *humidum radicale* as examples. The great complexity and at times inconsistency in the descriptions of these fundamental physiological concepts ultimately confront us with the question

[4] For example, Heinrich Haeser, *Lehrbuch der Geschichte der Medicin und der epidemischen Krankheiten*, 3rd edn, Jena 1875, vol. 1 (repr. Hildesheim 1971), p. 586; W. L. Dulière, "Avicenne, l'homme et sa signification dans la médecine européenne", in: *Le Flambeau* 34 (1951), pp. 484–505. Karl Eduard Rothschuh, *Konzepte der Medizin in Vergangenheit und Gegenwart*, Stuttgart 1978, pp. 199–201, also describes the *Canon* as a codification of Galen's teaching, does, however, note the deviation of Avicenna's theory of forms from the Aristotelian one. Alistair Cameron Crombie, "Avicenna's influence on the mediaeval scientific tradition, in: *Avicenna: scientist & philosopher. A millenary symposium*, ed. by George M. Wickens, London 1952, pp. 84–107, describes the *Canon* as innovative above all in Avicenna's methodical observation of pathological signs and the effects of medicines.

[5] Siraisi, Avicenna [see note 1] demonstrates this for selected fundamental natural-philosophical, physiological, and pathological questions.

[6] "His philosophy roughly represents the Aristotelian tradition as modified by neoplatonic influences and Muslim theology." (George Sarton, *Introduction to the history of science*, vol. 1: *From Homer to Omar Khayyam*), Baltimore 1927, p. 709, see also Constantin Sauter, *Avicennas Bearbeitung der aristotelischen Metaphysik*. Diss. Munich 1912; Dulière, Avicenne [see note 4] underscores the influence of mystical currents on Avicenna and his esoteric interests.

[7] See Ursula Weisser, "Die Harmonisierung antiker Zeugungstheorien im islamischen Kulturkreis und ihr Nachwirken im europäischen Mittelalter", in: Zimmermann et alii, Orientalische Kultur [see note 3], pp. 301–326, esp. pp. 306–307.

as to whether the enormous and lasting success of the *Canon* can, as is commonly assumed, be attributed simply to the work's systematicity and precision.

When reconstructing the *Canon's* fundamental medical concepts, a certain difficulty presents itself insofar as the work was conceived as a medical handbook or textbook. The author makes a point of renouncing any extensive theoretical explanations of natural-philosophical premises, stating that these would have no direct influence on the tasks of the physician and thus should be assigned to the domain of the philosopher.[8] Accordingly, the fundamental concepts are only touched on or are presented in a highly condensed manner and presented in a systematic structure, while thorough and comprehensive definitions are omitted. This was a difficulty—or challenge—naturally also encountered by the physicians of the Middle Ages and Renaissance, who were increasingly interested in philosophical matters.

The teachings about an inner, implanted heat as the source and basis of all life processes had been—along with the closely related ideas regarding an implanted spirit and a natural moisture—a central component of physiology and medicine since antiquity, without, however, having been elaborated as a truly coherent and generally accepted conception.[9] As will come to light here, this openness and incompleteness was continued in the *Canon*. The idea of a *calor innatus* in the *Canon* takes on explanatory functions in several contexts that often do not appear directly related, and thus it takes shape contextually in different, isolated passages that are not always easily reconciled.

[8] The argument is presented in particular detail in Avicenna, Canon, 1556 [see note 1], p. 6 (lib. 1, fen 1, doctr. 1, cap. 1).

[9] Brigitte Hoppe, *Biologie, Wissenschaft von der belebten Materie von der Antike zur Neuzeit*, Wiesbaden 1976; cf. Everett Mendelsohn, *Heat and life. The development of the theory of animal heat*, Cambridge 1964; Richard J. Durling, "The innate heat in Galen", in: *Medizinhistorisches Journal* 23 (1988), pp. 210–212; Friedrich Solmsen, "The vital heat, the inborn pneuma, and the aether", in: *Journal of Hellenic studies* 77 (1957), pp. 119–123.

Innate Heat and Natural Healing Power

There is no doubt about the central position of vital heat in the physiology of the *Canon*.[10] But, unlike with some of the other key concepts, one searches in vain for a clear definition of *calor innatus*. The closest one comes is in the characterization of the role it plays in natural healing:

> If, however, an extrinsic heat wants to destroy the equilibrium [of the qualities], the innate heat, which is found within, will be the stronger in resisting, to the point that the innate heat alone reins in and expels warm matter and corrupts its substance. After all, the innate heat is the instrument of nature that expels the harmfulness of the befalling heat for the reason that it moves the spirit so as to expel that heat, to remove its smoke, to dissolve it and to burn the matter of the same. And it also expels the harmfulness of the befalling coldness by exerting the opposite. [...] And it is also the innate heat that protects the natural humors so that extrinsic heat may not rule them. After all, when the innate heat is strong, nature can with its help proceed in the humors in the manner of maturation and digestion and can guard their health, and the humors are moved in its way or their motion according to the way of extrinsic heat is forbidden, which is why they do not rot. But if the heat is weak, nature tires and refrains from ruling over the humors due to the weakness of her instruments, because something stands between her and the humors, and the extrinsic heat finds that it [the innate heat] is not concerned with ruling. They [the humors] are then governed by the extrinsic heat, and it makes them obedient and moves them in a preternatural movement, and rot sets in. Innate heat, then, is the instrument of all faculties.[11]

[10] See, for example, Avicenna, Canon, 1556 [see note 1], p. 7 (lib. 1, fen 1, doctr. 3, cap. 1): "& uita quidem existit per caliditatem, & augmentum per humiditatem. Et etiam caliditas in humiditate existit, & ab ea nutritur." See also ibid., p. 705 (lib. 3, fen 21, tract. 1, cap. 2, on embryogenesis): "quòd non nutritur membrum uitale, dum non in ipso innititur uel fundatur uita cum calore naturali, spiritu uitali."

[11] Ibid., p. 84 (lib. 1 fen 2, doctr. 3, cap. 3): "Cum uerò extrinseca caliditas aequalitatem destruere uoluerit, calor innatus qui est intus, erit res fortior in resistendo, adeo ut uenena calida non refraenet, neque expellat, neque ipsorum corrumpat substantiam, nisi calor innatus. Calor nanque innatus naturae est instrumentum, quod caloris euenientis expellit nocumentum, propterea quod spiritum mouet ad ipsum expellendum & ad eius uaporem remouendum, & ad ipsum resoluendum, & ipsius materiam adurendum. [...] Et calor quidem innatus est ille, qui humores naturales tuetur, ne eis caliditates extraneae dominentur [.] Calor nanque innatus cum est fortis, potest natura ipso mediante in humorib[us] ministrare secundum modum maturandi & digerendi: & custodire eos secundum sanitatem: & mouentur humores secundum modum ministrationis ipsius, &

What we find here, then, is that the *Canon* ascribes a central role in bodily processes to the *calor innatus*. It is a decisive instrument of nature, though we do not learn whether this nature is individual and bound to the body or whether it transcends the individual and is all-encompassing.[12] Effectively, it guarantees the self-preservation of the organism as a whole because without it the ever-present "foreign" heat from the outside world would take over rule of the humors and rot would occur.[13] Its key role extends to an enabling of all faculties. This suggests that the faculties of the soul can only be actualized thanks to the mediation of the innate heat.

Innate Heat, Radical Moisture, and Natural Life Span

While the functional significance of the *calor innatus* can be roughly discerned from its connection with natural healing power and self-preservation, the question of its possible substrate, which, as a "calidum", would possess the quality of heat, remains open. All that are found are scattered references that are difficult to reconcile.

What emerges most clearly in this respect is the close relationship between the *calor innatus* and moisture, for example in the discussion of the temperaments in the different age groups and the causes of natural death—topics, in other words, that for their part exist in a close relationship to conceptions of natural healing power and self-preservation. With respect to the question of whether children or young adults ("juvenes") are of a hotter temperament, Avicenna joins Galen in positing that the

prohibentur moueri secundum modum ministrationis caloris extranei, quare non putrefiunt. Sed cum calor iste est debilis, fatigatur natura, & a regimine humorum abstinet propter instrumenti debilitatem, quod inter ipsam & humores est medium & stat: & calor extraneus inuenit ipsum non occupatum in regimine. Comprehenduntur ergo ab eo & facit eos obedientes, & motu extraneo ipsos mouet, & accidit putrefactio. Calor igitur innatus uirtutum omnium est instrumentum."

[12] In another work, Avicenna suggests that the inner heat and the nature of an individual are largely the same: "Les médecins emploient le mot tabi'a pour designer le tempérament, mizaj, la chaleur naturelle, al-harara al-gariziya, la disposition des membres, les mouvements, l'ame végetative" (cit. in Amélie-Marie Goichon, *Introduction a Avicenne. Son épitre des définitions. Traduction avec notes*, Paris 1933, p. 93, on the meaning of the term "tabi'a" ("individual nature").

[13] Later authors, sometimes claimed that innate heat and the natural healing power were the same; see Max Neuburger, *Die Lehre von der Heilkraft der Natur im Wandel der Zeiten*, Stuttgart 1926.

heat is ultimately the same with both: in children there is a greater quantity but the heat is less acrid, while adults have a smaller quantity that is more acrid, however.[14] Initially, he states, the moisture is found in such a great quantity that it both retains the heat and accomplishes growth. Ultimately, however, when growth has reached an advanced state, it can only fulfill one of the two tasks. And this can only be the preservation of heat, which for its part is the basis of growth.[15]

Moisture appears here as the material basis of growth upon which the soul or nature, following God's will, acts, with the *calor innatus* as an instrument. This occurs without an explicit postulation of a complexional unification of heat and moisture. Indeed, the ability of moisture—conceived as solely passive but also as matter, thinking along Aristotelian lines,—to "tend to" or "preserve" the heat suggests that moisture has its own complexional character as a "humidum".[16]

Following this, the human life span and natural death are directly linked with the consumption of moisture as the substance of the *calor innatus*.

> In addition, one must know that heat after the age of maturity begins to decrease: this is because the surrounding air dries out its matter, which is moisture. The innate heat in the body, too, contributes to this, as does the exertion connected to movements of the body and the soul, which are

[14] Avicenna, Canon, 1556 [see note 1], p. 10 (lib. 1, fen 1, doctr. 3, cap. 3): "Et huius [Galeni] quidem declaratio [...] est, ut opinemur unam solam caliditatem in mensura: aut corpus subtile, calidum, unum in quantitate & qualitate in substantiam humidam multam semel sparsum esse, sicut aqua: & alia uice sparsum esse in substantiam siccam & paucam, sicut lapis, & cum sic fuerit, inueniemus caliditatem aqueam maioris quantitatis, & mollioris qualitatis: & caliditatem lapideam minoris quantitatis, & acutioris qualitatis: & ita debes agnoscere caliditatem quae est in pueris & iuuenibus." Cf. Galen, "Peri kraseon", book 2, cap. 3, in: idem, *Opera omnia*, 20 vols. Ed. by Carl G. Kühn, Leipzig 1821 –1833, vol. 1, pp. 577–598.

[15] Avicenna, Canon, 1556 [see note 1], pp. 10–11 (lib. 1, fen 1, doctr. 3, cap. 3): "erat igitur prius humiditas in tanta quantitate, quod duas res facere poterat, custodire uidelicet calorem, & superesse etiam ad augmentum faciendum, ad ultimum uerò ad hoc deueniet, quod neutra duarum rerum efficere ualebit. oportet ergo ut in medio talis existat, quod possit unam rem efficere, & non alteram."

[16] Ibid., p. 11: "Humiditas nanque augmenti est materia. Et materia non patitur per se, neque per se creatun immò apud operationem uirtutis in eam agentis. uirtus autem hoc agens, est anima, aut natura praecepto Dei, neque agit nisi cum instrumento: & illud est calor innatus."

inevitable in life, and the inability of nature to always resist this. After all, all physical faculties are finite [...].[17]

And so, "by necessity, the moisture will end and the heat extinguished, all the more so because the required nourishing substance is lacking", and the increasingly insufficient digestion exposes the inner heat to the effects of a preternatural humor (*humor extraneus*), which suffocates it through its slimy constitution and counteracts it through its cold quality.[18] "And this is natural death, which is predetermined as the endpoint of every individual, the time when it no longer lies within his power to conserve his own moisture, which depends on his original mixture".[19]

A similarly close relationship between the *calor innatus* and moisture—although here again, the relation is not clearly explained—is described in the chapter "De causis sanitatis & aegritudinis, & de neceßitate mortis".[20] Puzzling as it is in the context of an interaction of elemental qualities, the aforementioned reference to the suffocation of the *calor innatus* by preternatural humors is explained here with a metaphor that has its roots in Galen: *calor innatus* is likened to the flame of an oil lamp.[21] Different extrinsic and intrinsic influences as well as the *calor innatus* itself all have

[17] Ibid.: "Praeterea sciendum est, quod caliditas post tempus consistendi minui incipit: propterea quod aer circundans materiam eius quae est humiditas, exiccat. Calor quoque innatus, qui est intus, ad hoc adiuuat: & labor proueniens ex motionibus corporeis & animalibus, quae necessariae sunt in uita, & defectus naturae ad resistendum illi semper. Omnes enim uirtutes corporeae sunt finitae." The Aristoteliaen idea that the "virtutes" are by their very nature finite, seems per se difficult to reconcile with the Galenic theory of qualities.

[18] Ibid.: "oportet necessariò finiatur humiditas, & calor extinguatur, & praecipue cum ad eius extinctionem adiuuet praeter materiei defectum, alia causa. Et illud est humor extraneus, qui accidit assidue, propterea quod cibus priuatur digestione, & adiuuat ad extinguendum ipsum duobus modis: uno, suffocando & tegendo: alio, qualitati obuiando. Hic enim humor est phlegmaticus frigidus."

[19] Ibid.: "Et haec est mors naturalis, cuique indiuiduo destinata, secundum suam primam complexionem, usque ad terminum quem in sua potentia habet ad suam humiditatem conseruandam."

[20] Ibid., pp. 104–106, lib. 1, fen 3, single chapter preceding doctr. 1.

[21] On the history of the lamp metaphor since Antiquity see Peter H. Niebyl, "Old age, fever, and the lamp metaphor", in: Journal of the history of medicine 26 (1971), pp. 351–368.

the effect they promote a drying out. According to Aristotle, this ultimately lethal process of drying out has to be considered indispensable to the preservation of the body and its natural functions while alive:

> And so, we need this dryness that befalls us. Because, after all, we consist of moisture from beginning to end: for this reason, it is certainly necessary for heat to overcome it, as it [the heat] would itself be suffocated otherwise: it doubtlessly acts upon it [the moisture] perpetually, and works to dry it out more and more. [...] The inborn heat is, in other words, accidentally the cause of its own extinguishment, for the reason that it consumes its substance.[22]

Insofar as the *calor innatus*, due to its weakness, is ultimately less and less able to supply matter as a replacement for the exhausted moisture, instead accumulating poorly assimilated humors, this effect is strengthened.

> The weakness of the heat increases, because dryness in the substance of the body parts is predominant and because the natural moisture is diminished, which is like a substance, like the oil of a lamp. After all, the lamp has two moistures, water and oil, and the one constitutes it while the other extinguishes it. Similarly, the inborn heat is constituted in natural moisture and is suffocated or extinguished by the foreign, with the increase of the foreign moisture which arises from the weakness of digestion and which relates to it like a watery liquid to the flame of a lamp. And when the dryness is complete, the natural heat is extinguished and natural death occurs.[23]

[22] Avicenna, Canon, 1556 [see note 1], p. 105 (lib. 1, fen 3, single chapter preceding doctr. 1): "Haec igitur siccitas, quae nobis accidit, res est nobis necessaria. Nos siquidem a principio rei in ultima consistimus humiditate: unde proculdubio oportet ut calor eam superet, quoniam si non, in ipsa suffocaretur: procedit igitur proculdubio in ipsam assidue agendo, & ipsam semper exiccando incedit. [...] Calor igitur innatus est causa extinguendi seipsum accidentaliter propterea quod fit causa suiipsius materiae consumendi."

[23] Ibid.: "Augmentatur igitur caloris debilitas, propterea quod siccitas in membrorum substantia dominatur: & propter naturalis humiditatis diminuitionem, quae est sicut materia, & sicut oleum lampadis. Lampas nanque duas habet humiditates, aquam & oleum, cum una quarum consistit, & cum altera extinguitur. Similiter calor innatus cum humiditate consistit naturali, & suffocatur seu extinguitur cum extranea, & cum augmento extraneae humiditatis, quae debilitate digerendi: & est sicut aquea humiditas flammae lampadis. Et cum siccitas ad complementum accedit, calor innatus extinguitur, & mors subsequitur naturalis."

In other words, the loss of moisture, which is deadly in the long run, is to a certain degree a necessary accidental concomitant effect of vital processes. The inner heat consumes its own substrate and is increasingly threatened by foreign moisture that extinguishes the flame.

Peter H. Niebyl has pointed out the originality and lasting impact of this application of the traditional metaphor of the lamp. Generations of physicians, from Roger Bacon to Jean Fernel to Thomas Willis, took recourse in their accounts of the physiology of aging to the *Canon*.[24] The *Canon's* descriptions of the natural aging process by no means make it easier, however, to understand the function and significance of the *calor innatus* within the physiology of the *Canon*. Not only is the *calor innatus* supposed to possess extraordinary, supra-elementary abilities, its moist substrate, too, is described as a very special humor that can only be incompletely renewed through nourishment. So where does the *calor innatus* come from?

The lamp metaphor also finds what was in Avicenna's time a traditional application in another explanatory context in the *Canon*, namely in the discussion of "hectic fever". Here, the three stages of hectic fever are ascribed to those three of four secondary moistures that are found outside the veins.[25] The first, dew-like moisture of the three moistures associated with hectic fever may be transformed into nutrition and is able to moisten the body parts when a drying out threatens to occur. The second moisture, which is close to solidifying, feeds the parts directly but, with respect to its nature has not been fully assimilated by them. The third moisture, finally, has its beginnings ["principium"] in the semen and so, unlike the other three, it is not owed to the continuous processes of digestion; from birth, it is found in the simple parts, guaranteeing their cohesion.[26]

[24] Niebyl, Old age [see note 21]; Marsilio Ficino, a key figure in the history of medical dietetics whom Niebyl does not mention, is also relevant here, especially book 2 of his *De vita*, entitled "De vita longa" (Marsilio Ficino, *Three books on life*. Ed. and transl. by Carol V. Kaske and John R. Clark, Binghampton, NY, 1989) still in the seventeenth century, Charles Valois, in his commentary on the *Canon*, rejected this view, which he considered heretical (Siraisi, Avicenna [see note 1], p. 213).

[25] The first of these four secondary moistures is located in the openings of the small vessels and imbibes from there the surrounding parts [see note 26].

[26] Avicenna, Canon, 1556 [see note 1], p. 801 (lib. 4, fen 1, tract. 2, cap. 1); cf. ibid., p. 12 (lib. 1, fen 1, doctr. 4, cap. 1): "Vna est humiditas in foraminibus extremitatum paruarum uenarum contenta, membris simplicibus propinquarum imbibentium ea, uel a quib[us] ipsa membra imbibunt. Alia est humor, per omnia simplicia transiens membra:

These three secondary moistures associated with the three stages of hectic fever are then compared with the oil in the lamp's oil receptacle, the oil that has saturated the wick, and the liquid that holds together the wick itself:

> And the first-named moisture corresponds to the oil that has been poured into the lamp; the second-named to the oil that has been drawn into the body of the wick, and the third-named moisture is the moisture through which the parts of the the cotton from which the wick is made are held together.[27]

In the case of hectic fever, the heat first consumes the moisture of the first kind in the body parts and above all in the heart, then that of the second as the clinical picture of marasmus emerges, and ultimately the heat begins to consume the third kind of moisture and "the flame begins to burn the body of the wick and its elemental moistures".[28]

Thanks to its introduction in the *Canon*, the term ***humidum radicale*** was to become a key concept in the discussion of life processes. The equation of the ***humiditates radicales*** with the moist substrate of the ***calor innatus***, which became common in subsequent times, seems to follow from the text of the *Canon* in the eyes of modern interpreters such as

sicut ros qui in nutrimentum conuerti est aptus, cum corpus nutrimento caret, & ut membra humectet, cum aliqua causa fortis motus, aut alia ea exiccauerit. Tertia est humiditas, quae in tempore proximo fuit coagulata. Et est nutrimentum, quod in substantia membrorum ex parte complexionis & similitudinis conuersum est, sed ex parte essentiae completae nondum conuersum fuit. Quarta est humiditas quae est intus in membris simplicibus a principio natiuitatis, per quam eorum continuitas existit: cuius principium est ex spermate." I therefore do not agree with Michael R. McVaugh when he equates the fourth of these secondary moistures with the "final humoral stage in the conversion of food to bodily substance" (Michael R. McVaugh, "The 'humidum radicale' in thirteenth-century medicine", in: *Traditio* 30 (1974), pp. 259–283, cit. p. 266). It is precisely not a product of the conversion of food.

27 Avicenna, Canon, 1556 [see note 1], p. 801 (lib. 4, fen 1, tract. 3, cap. 1): "Et similitudo humiditatum primarum est oleum lucernae effusum in lucerna. Et similitudo secundarum est oleum imbibitum in corpore lychinij. Et similitudo tertiae humiditatis, est humiditas qua continuantur partes cotti, de quo factus est lychinius."

28 Ibid., pp. 801–802 (erroneously paginated as p. 856): "incipit finire humiditates, quae sunt diuisionis tertiae, sicut incipit flamma adurere corpus lychinij, & humiditates eius radicales".

Thomas S. Hall and Michael R. McVaugh as well.[29] Yet a close reading leads to a significant degree of doubt in this regard.

We can establish first of all that, when the aging process is discussed, mention is made of a *calor innatus*, while with hectic fever, it is a *calor extraneus*. What applies to the effect of feverish heat does not necessarily apply to innate heat. In *De marasmo*, Galen voiced his opposition to the idea that the *calor innatus* has a fire-like character,[30] and the *Canon*, too, insists on a fundamental distinction between the two kinds of heat.[31] It cannot be concluded in other words that the two consume the same moisture as fuel despite all differences, though it cannot be ruled out either.

However, the crucial point here is not just that the lamp image is applied to two different bodily processes, aging and hectic fever, but that the metaphor works differently in the two contexts. When the aging process is described, it is enough to attribute the cause of death to an evaporation or dilution of a liquid likened to lamp oil that is thought to feed the heat. In the case of hectic fever, the consumption of the lamp oil only constitutes the first step, and death follows the burning up of the wick itself. Does it truly make sense, then, to conclude that the processes are analogous and above all that the two fuels are one and the same?[32] This equation of the two processes that later authors have postulated is articulated nowhere in detail. Rather, this conclusion seems to arise from an associative chain that is only loosely expressed in the *Canon* between sperm and *humiditates radicales* on the one hand and between sperm and the *calor innatus* on the other.

There is plenty of evidence of the close connection between sperm and the *calor innatus* in the *Canon*, even though no mention is initially made of the *calor innatus* in the discussion of the procreation

[29] McVaugh, The humidum radicale [see note 26]; Thomas S. Hall, "Life, death, and the radical moisture", in: *Clio medica* 6 (1971), pp. 3–23.

[30] Galen, "De marcore", in: idem, Opera [see note 14], vol. 7, pp. 665–704; cf. Theoharis C. Theoharides, "Galen on marasmus", in: *Journal of the history of medicine* 26 (1971), pp. 369–390, with an English translation.

[31] Avicenna, Canon, 1556 [see note 1], p. 753 (lib. 4, fen 1, tract. 1, cap. 1).

[32] According to Per-Gunnar Ottosson, *Scholastic medicine and philosophy. A study of commentaries on Galen's 'Tegni' (ca. 1300–1450)*, Naples 1984, note on p. 252, Torrigiani and Jacopo da Forlì did not use the term "humidum radicale" in their discussion of the "calor innatus".

process and embryogenesis. The *virtus informativa* of the sperm appears to be endowed with the immediate ability to generate a "frothiness" (*spumositas*) of the sperm and to move the animal, natural, and vital spirits in the sperm to toward the location where the different organs are formed.[33] The first organ to be formed is the heart, so that it can serve in the first place as the source of natural heat and vital spirit for the other body parts.[34] However, what is suggested, despite the absence of an explicit mention of innate heat, by the processes of movement in these very first stages of embryogenesis, is confirmed in a different place, namely in the discussion of sterility: "And when the male member is too long, the distance the sperm travels is lengthened, which is why its innate heat is already broken when it reaches the uterus, and so for this reason it usually does not generate".[35] As it turn out that the *virtus formativa* of the sperm requires as its instrument the innate heat, it follows that it is already present in the sperm.

The relationship the sperm has to the *humiditates radicales* is less clear. Nowhere in the *Canon* is it stated unambiguously that the latter are already present in the sperm, which might render them identical with the moist substrate of the innate heat. All that it says is that the *humiditates radicales*, as described in the quotation cited above, have their origin in the sperm. An additional difficulty presents itself insofar as the sperm for its part does not arise from the *humiditates radicales* in the body parts, but rather from the process of food concoction:

> The sperm is a surplus of the fourth digestion which takes place, when the food that is distributed to the body parts, sweating out from the veins, after the third digestion is completed. And it consists of the most noble of the moistures which is close to the time of solidification and from which

[33] Avicenna, Canon, 1556 [see note 1], p. 795 (lib. 3, fen 21, tract. 1, cap. 2): "Cum matrix colligitur super sperma, tunc prima dispositionum quae aduenit illic, est spumositas spermatis & est ex operatione uirtutis formativae. Et ueritas quidem dispositionis ipsius spumositatis est ex hoc quod uirtus informatiua mouet illud quod est in spermate ex spiritu animali naturali & uitali ad mineram uniuscuiusque eorum".

[34] Ibid.

[35] Ibid., p. 690 (lib. 3, fen 20, tract. 2, cap. 13): "Et cum prolongatur uirga ualde, prolongatur spacium motus spermatis: quare uenit ad matricem calore eius innato iam fracto, non ergo generat secundum plurimum."

the principal body parts are nourished, such as the veins, arteries, and the like.[36]

Spermatogenesis, in other words, does not correspond to a mere transmission of the *humiditates radicales*, which from birth onward, and across the generations, guarantee the cohesion of the body parts. Sperm is always created anew from the third secondary moisture which, unlike the *humiditates radicales*, is generated from the concoction of food.[37]

The explanations of these matters in the *Canon* are anything but clear. The possibility remains that the *humiditates radicales* and the moist substrate of the *calor innatus* coexist in the sperm and it cannot even be ruled out entirely that they are indeed identical. This would, however, create new and irresolvable contradictions. After all, if the sperm—containing the moist substrate of the *calor innatus*, which is measured out to last the length of an entire human life—is owed to the concoction of food, the question arises why the moist substrate of the *calor innatus* cannot be sufficiently renewed in the concoction processes of the individual in question. If it could be, Avicenna's novel understanding of the naturally limited life span as the necessary consequence of physiological life processes would be rendered invalid.

[36] Ibid., p. 688 (lib. 3, fen 19, tract. 2, cap. 3):"Sperma est superfluitas digestionis quartae, quae fit cum dispertitur cibus in membris, resudando a uenis, tertia digestione iam expleta. Et est de summa humiditatis proximae tempori coagulationis, ex qua nutriuntur membra radicalia, utpote uenae, arteriae, & similia illis"; see also the account of the third moisture which was associated with the second stage of hectic fevers, ibid., p. 801 (lib. 4, fen 1, tract. 3, cap. 1): "Et de illis sunt humiditates proximi temporis coagulationi: et sunt humiditates, quae fiunt actu nutrimentum ueruntamen tempus fluxbilitatis [sic!] earum propinquum est quare sunt non coagulatae."

[37] The fourth digestion or stage of concoction is not linked to the fourth secondary moisture. We are dealing here with two different foursomes. The third and fourth digestion, following those in the stomach and the liver, take place in the veins and in the various parts of the body. See ibid., p. 15 (lib. 1, fen 1, doctr. 4, cap. 2): "Debes autem scire, quod sanguis & quod cum eo labitur, tertiam habet in uenis digestionem: & cum per membra diuiditur, ut unumquodque suum recipiat nutrimentum, habet digestionem quartam." How exactly spermatogenesis is supposed to occur, remains unclear. Presumably, the idea was that of a pangenetic origin of spermatic matter throughout the body and its parts, which was then further reinfed in the testicles; cf. ibid., p. 687 (lib. 3, fen 19, tract. 3—erroneously numbered tract. 1—, cap. 1): "Svblimis Deus creauit duos testiculos [...] qui sunt membra principalia in quib. generatur sperma ex humiditate delata ad eos in uenis: quae est quasi superfluitas cibi quarti in corpore toto. Et est sang. magis digestus, & magis subtilis." On the history of theories of conception see Erna Lesky, *Die Zeugungs- und Vererbungslehren der Antike und ihr Nachwirken*, Wiesbaden 1951.

Innate Heat Between Primary Quality and Complexio

How can the indications that the *calor innatus* has an elementary nature, a complexional constitution as a *calidum innatum* with heat as its predominant quality and moisture as a substrate be reconciled with its complex faculties and difficulties in renewing itself from food? Are the physics of elemental qualities being transcended here? Are we perhaps even dealing with stellar matter? Or, considering the complex functions of the *calor innatus* in the human organism, can it nevertheless be understood as an elemental quality in the end—one that owes its particular effects at best to the guiding influence of the superior faculties of the soul? Let us first consider its role in the different functional contexts.

The elemental-qualitative character of the *calor innatus* becomes readily understandable in the description of the nutritive faculty, the *virtus nutritiva*, and its four auxiliary faculties, the attractive, retentive, digestive, and expulsive faculties.[38] Thanks to heat, and under the influence of the formal cause, the *virtus naturalis* is able to work on food, but also on potentially harmful excess matter, either concocting it into a new *complexio* that is suitable to the body, or else, with the motive force of the heat, discharging it from the body if it is matter that cannot be transformed as described.[39] Not only this, but the aptitude for self-preservation—the nature of the body—can also make use of the opposing elemental quality of the *calor innatus* to combat the harmful influence of external cold.[40]

Yet, these descriptions are outweighed by the much more numerous passages that cannot be adequately reconciled with the idea that the *calor innatus* has a merely elemental character, even when it is seen as an instrument of a higher entity or when it is subject to the influence of a formal cause. An example here is the idea sketched out above that foreign heat with its hostile effect on the body can be warded off by the body's own *calor innatus*. Only on the surface does this appear to refer to the effect that heat—as an active, motive primary quality—may have on the humors.

[38] Avicenna, Canon, 1556 [see note 1], p. 48 (lib. 1, fen 1, doctr. 6, cap. 2); the "virtus nutritiva" and the "virtus crescitiva" together formed the "virtus naturalis".

[39] Ibid.

[40] Ibid., p. 84 (lib. 1, fen 2, doctr. 3, cap. 3): "Et frigidi etiam aduenientis nocumentum per contrarium expellit."

On an elemental-qualitative level, the interaction of the two types of heat ultimately remains opaque. The image of the dominion that is wrested from the *calor innatus* by external foreign heat so as to direct the movement of the humors can essentially only be attributed exclusively to the effect of elemental heat if one denies the obvious regularity of pathological processes. While for *calor innatus*, one could still point to an instrumentalization of its elemental action by higher entities in the body, that is the *virtutes* of the soul, this is not possible for its analogue, the *calor extraneus*, which is likewise endowed with complex ruling faculties.

Furthermore, an unhealthy excess of *calor innatus* is as good as ruled out, though this would have to be the case for pure, elemental heat. Rather, in the context of the teachings on the pulse, a stronger *calor innatus* is identified with a strengthening of the natural functions.[41] Also, it is difficult to explain in terms of primary qualities the beneficial effect that is ascribed—making reference to the Hippocratic aphorism I, 15—to a concentration of *calor innatus* in the body during the wintertime.[42] Incidentally, the imagery of the flame as a manifestation of heat does point beyond the limits of an elemental quality: it spreads light, can be propagated without being diminished itself, and cannot be overpowered by the qualitatively opposing cold, but rather can be suffocated or may die due to lacking fuel.

In no place, however, does one find an unambiguous definition of the constitution of the *calor innatus* as supra-elementary, complexional, and endowed with a special form—not even in those places where such a constitution becomes almost palpable, namely when the specific faculties of heat are distinguished from the inferior faculties of cold. The passage in question states that the cold in the body is only able to fight its opposite, but not to protect the body from external cold. Ultimately, it tends to be harmful, and could be useful at most "per accidens".[43] Here, innate heat clearly assumed a special status, one that cannot be simply owed to

[41] Ibid., p. 92 (lib. 1, fen 2, doctr. 3 cap. 8).

[42] Ibid., p. 129 (lib. 1, fen 3, doctr. 5, cap. 1): "Calor enim innatus, qui est regitiuus; in hyeme uehementer fit fortis propterea quod a resolutione salvatur, & coadunatione unitur. Omnes praeterea naturales uirtutes suas melius perficiunt operationes." Cf. Hippokrates, "Aphorisms" in: idem, *Œuvres complètes*. Ed. by Émile Littré, Paris 1839–1861, vol. 4, p. 466 (aphorism I, 15).

[43] Avicenna, Canon, 1556 [see note 1], p. 84 (lib. 1, fen 2, doctr. 3, cap. 3): "Hanc uerò proprietatem non habet frigiditas: non enim impugnat neque refraenat nisi aduenientem caliditatem per contrarium tantum, quia non impugnat aduenientem frigiditatem.

the action of the *virtutes*. The source of this special characteristic is made explicit: "one speaks of innate heat but not of innate cold and that which is allocated to heat from the *colcodea* is not allocated to cold".[44]

As we shall see, the special faculties of innate heat are not owed to the faculties of the soul exclusively. Rather, they originate with an external entity, with the mysterious *colcodea* of Renaissance writing, which is usually rendered as "dator formarum" in Latin. The *Arabicorum nominum antiqua expositio* that was added to Rinio's edition traces the term to the Persian "colcodeo", meaning, more or less, "dator vitae".[45] In the framework of Avicenna's emanatory metaphysics, the *colcodea* corresponds to the tenth intelligence that controls the fate of the sublunar world by acting upon matter that is so disposed.[46] Accordingly, the particularities, the complex faculties of innate heat are not exclusively owed to an elemental mixture, but also to an external, celestial source of forms.

[...] Sed frigiditas est ei [the innate heat, M.S.] inconueniens: & neque prodest, nisi accidentaliter."

44 Ibid.: "dicitur calor innatus, & non dicitur frigiditas innata, neque proportionatur ad frigiditatem illud, quod ex colcodea corporis ad caliditatem proportionatur."

45 "Arabicorum nominum antiqua expositio", in: Avicenna, Canon, 1556 [see note 1], no page numbers: "Colcodea persicum nomen est deriuatum a colcodeo qui est dator uitae." on the roots and uses of the term, which has astrological origins and was frequently distorted in medieval and Renaissance writing, see Bruno Nardi, "Origine dell'anima umana", in: idem, *Studi su Pietro Pomponazzi*, Florence 1965, pp. 231–246, note on pp. 234–235; and C. A. Nallino, "La 'Colcodea' d'Avicenna e T. Campanella", in: *Giornale critico della filosofia italiana* 6 (1925), pp. 84–91. According to Nardi, the widespread use of the term in the fourteenth and fifteenth centuries was owed to Pietro d'Abano and his "Conciliator". He seems to have found it in Avicenna's "Metaphysica" and used the form "alcochoden". We find "colchodea" only in Argilageus' revised 1483-edition of the "Conciliator". Unware of the "colchodea" in the translation of the *Canon*, Nardi expressed his astonishment at this change but presumably Argilaeus drew on this translation in 1483. Ultimately, the form "colchodea" and the defintion of the term in the *Antiqua expositio* thus probably originated from Gerhard of Cremona's translation. A systematic survey of extant manuscripts of the Latin *Canon* could offer supporting evidence for the assumption. Two manuscripts of the *Canon* which I have seen in the Bayerische Staatsbibliothek in Munich (Cod. lat. 5353 and Cod. lat 13017) do in fact use "cholchodeia" and "colchodeia" respectively; moreover, a glossary in Cod. lat. 13017 traces the term "colchodeia" back to the Persian term "colcodeio".

46 Nallino, La 'Colcodea' [see note 47]; Soheil M. Afhan, *Avicenna. His life and works*, London 1958, pp. 126–135.

Innate Heat and Innate Spirit

To further clarify the position of *calor innatus*, it makes sense at this juncture to investigate its connection to that unique substance in the body and in sperm whose matter Aristotle had compared to that of heavenly bodies: the *spiritus innatus*. In it, the medical tradition linked to combined to varying degrees soul-like features in the sense of highly complex, purposeful faculties, and ideas of an air-like material nature, which was believed to become evident in the frothy constitution of sperm.[47]

As the quotations cited above indicate, the doctrine of *spiritus* plays a significant role for Avicenna and was without question closely linked to his teaching about the *calor innatus*. In one place, for instance, he states that "The heart and the *spiritus* are truly the foundation of life".[48] The terminology, however, often makes it difficult to arrive at a closer definition. Only in very rare instances do we find use of the term *spiritus innatus*, which would seem a most likely potential counterpart to the *calor innatus*. Rather, the general term *spiritus* is used much more frequently, or the text is oriented on the traditional division of natural, vital, and animal *spiritus*, which were thought to already develop their effectiveness in the sperm.[49] The only explicit and thorough acknowledgment of the role of a *spiritus innatus* in the *Canon*, in the discussion of respiratory processes, does not clarify the matter conclusively either. It states that when air is drawn in toward the heart via the lungs and pores, two purposes are fulfilled at once, namely a renewal of the *spiritus* through the intake of air as one of its elements on the one hand, and on the other a cooling and airing of the *spiritus innatus*.[50] For, the external

[47] Werner W. Jaeger, "Das Pneuma im Lykeion", *Hermes* 48 (1913), pp. 29–74; Arthur L. Peck, "The connate pneuma. An essential factor in Aristotle's solutions to the problems of reproduction and sensation, in: *Science, medicine, and history*, ed. by E. Ashworth Underwood, Oxford 1953, vol. 1, pp. 111–121; Owsei Temkin, "On Galen's pneumatology", in: *Gesnerus* 8 (1951), pp. 180–189; on the theory of spirits in medieval medicine see James J. Bono, "Medical spirits and the medieval language of life", in: *Traditio* 40 (1985), pp. 91–130.

[48] Avicenna, Canon, 1556 [see note 1], p. 7 (lib. 1, fen 1, doctr. 3, cap. 1): "Vitae etenim principium, sunt cor & spiritus: quae quidem ambo uehementer calida existunt…".

[49] Ibid., p. 705 (lib. 3, fen 21, tract. 1, cap. 2).

[50] Ibid., p. 57 (lib. 1, fen 2, doctr. 2, cap. 1): "Aer est elementum nostrorum corporum & spirituum: et praeter hoc quod est element[um] nostrorum corporum & spirituum, est etiam quid successiue nostris adueniens spiritibus: & est causa melioriationis eorum, non tantum sicut elementum, sed etiam sicut faciens, scilicet temperans."

air was, compared to the hot *complexio* of the *spiritus innatus*, colder by far, especially when the *spiritus innatus* was compressed.[51]

> Whenever, air reaches it [the *spiritus innatus*], collides and mixes with it, and keeps it from changing into a compressed or stifling fieriness from which its *complexio* would deteriorate, through which it would lose the ability to receive the impressions of the soul in that which is the cause of life, and it [the air] also keeps its [the *spiritusinnatus*] vaporous, moist substance from dissolving.[52]

It is clear, then, how the *spiritus innatus* is ascribed a complexional constitution while its airy character is emphasized as well. Insofar as the air possesses warm and moist qualities, this would be entirely reconcilable with the idea that the moist substrate of the *calor innatus*, which was likewise always threatened by dissolution and evaporation, could be identical with the *spiritus innatus*. But then the reason why the substrate could only be renewed to a limited extent would remain unclear.

Spiritus is described slightly differently in a detailed definition found in another of Avicenna's medical treatises that was often published alongside the *Canon* in later printed editions, *De medicinis cordialibus*, translated into Latin by Arnaldus de Villa Nova[53]:

[51] Ibid.: "Aer autem qui nos circundat, ualde est frigidus comparatione complexionis, spiritus innati, quanto plus complexionis eius euenientis propter coarctationem."

[52] Ibid., p. 57 (lib. 1, fen 2, doctr. 2, cap. 2) "Cumque aer ad ipsum peruenit, impellit ipsum, & miscetur cum eo, & prohibet ipsum ne ad igneitatem conuertatur coarctatiuam uel suffocatiuam, ex qua ipse ad malitiam peruenit complexionis, per quam remouetur a praeparatione recipiendi impressiones animales in eo qui est causa uitae, & prohibet ne resoluatur ipsiusmet substantia uaporalis humida."

[53] Avicenna, "De medicinis cordialibus", in: idem, Canon, 1556 [see note 1], pp. 1040–1058; this is a very striking text which differs markedly in its termionology and content from the *Canon* and shows neoplatonic influences but has so far attracted little attention. An exception is Shifaulmulk Hakim Abdul Latif, "Introduction to "Heart drugs", a brilliant work of research by Avicenna, in: Wickens, Avicenna [see note 4], pp. 245–254; his claim, however, that this text could "revolutionize" our modern ideas about treating diseases of the heart and that Avicenna was the first to show the close relationship between psychology and medicine makes is unconvincing. Scholars have not expressed doubts on Avicenna's authorship; see, for example, Lucien Leclerc, *Histoire de la médecine arabe*, Paris 1876, vol. 1; Donald Campbell, *Arabian medicine and its influences on the Middle Ages*, London 1922, vol. 2, p. 4, suggests 1282 as the year of Arnaldo's translation. Others have edited the text under the title "De viribus cordis" (see Ludwig Choulant, *Handbuch der Bücherkunde für die ältere Medicin*, 2nd edn, Leipzig 1841 (repr. Graz

> In summary, the spirit is therefore something that is formed from a mixture of the elements and tends toward a similarity to the heavenly bodies. And for this reason, it is said of spiritus that it is a luminous substance, which is why one says of the visual spirit that it is ray and light. For this reason, the soul is joyful upon perceiving light, and becomes somber in the dark, because the former is indeed in keeping with the nature or foundation of the same, while the latter stands contrary to it.[54]

Regarding the origin of this spirit, Avicenna concluded:

> God created from the cavities of the heart the left one, so that it would serve as the receptacle of the spirit and its place of generation. [...] He therefore created first the spiritus as an appendage of the faculties of the soul and the second emanation in the body parts by means of the same. He also created the spirit from the refined parts of the humors and from their vapors.[55]

So here the spirit appears without any further differentiation to be the immediate instrument of the soul; there is no mention of the role of the innate heat as would be expected at this juncture from the description in the *Canon*. One would tend to understand this, too, as pointing to an exchangeability of the two by and large, whereby the term "spirit" would refer more to the material aspect and "innate heat" more to the qualitative aspect, to that which communicates the faculties. The spirit is based in the substance of the humors, yet, due to its similarity to light, it exhibits characteristics that by far transcend the characteristics of the elements.

1956), pp. 362–367). The different titles reflect the two major themes which the treatise discusses.

54 Avicenna, De medicinis cordialibus [see note 53], p. 1040 (tract. 1, cap. 2): "& spiritus quidem in summa est res generata ex commistione elementorum, uergens in similitudinem coelestium corporum. Et propterea iudicatur de spiritu, quod sit substantia luminosa: propter quod dicitur de uisiuo spiritu, quod sit radius atque lux. Et exinde anima gaudet ex intuitu lucis, & tristatur in tenebris: quoniam illud est conforme naturae uel principio ipsius, & hoc est ei contrarium."

55 Ibid., p. 1040 (tract. 1, cap. 1): "CREAVIT Deus ex concauitatibus cordis sinistram, ut esset armarium spiritus, et minera generationis illius. Creauitque spiritum latorem uirtutum animae in membra animalium [...]. Creauit itaque primum appendiculum uirium animae spiritum & manationem secundam in membris corporis mediante ipso. Produxit etiam spiritum ex subtili parte humorum, & ex uaporibus ipsorum."

Innate Heat and the Doctrine of Forms

Attributing the special powers of innate heat to the *colcodea*, a celestial giver of forms, has far-reaching theoretical consequences. It implies a rejection of the "eductio formarum ex potential materiae", of attempts to see all natural processes—all becoming and ceasing to be—as exclusively due to the interaction of the primary qualities in the *complexio*. For this reason, I will consider the role of innate heat in the *Canon's* doctrine of forms a little more closely in this last section.

The *Canon* ascribes to every living thing and every body part a particular *complexio*, as given by God.[56] In the complexional union, the elemental particles with their forms remain intact, however. "The elements are simple bodies, and they are the first parts of the human body and of the other bodies that can in no way be divided into bodies that have other forms: from their mixing, the different kinds of that which has become are created."[57] What is at work here, in other words, is a corpuscular conception of the process of mixing that assumes the existence of very small particles that are endowed with certain forms.[58] At the level of qualities, the *commistio* of the elemental particles corresponds to the conception of the nature of the *complexio*[59]:

> Complexio is a quality that emerges from the mutual acting and suffering of contrary qualities in the elements. Their parts are reduced to such minuteness that each one of them touches the most and is changed the

[56] Avicenna, Canon, 1556 [see note 1], p. 9 (lib. 1, fen 1, doctr. 3, cap. 2): "Dico, quod Deus cuique animali, & cuique membro donauit ex complexione quod magis mereatur". As in the case of the spirit, this would not seem to exclude a mediating role of lower intelligences in the act of creation, even if they are not explicitly mentioned.

[57] Ibid., p. 6 (lib. 1, fen 1, doctr. 2, single chapter): "Elementa sunt corpora simplicia, & sunt partes primae corporis humani, & aliorum quae in corpora diuersarum formarum diuidi minime possunt ex quorum commistione species diuersae generatorum fiunt."

[58] In his metaphysics, Avicenna rejected the classical atomistic theories which were widespread in Islamic theories of creation before him; cf. Soheil M Afnan, *Avicenna. His life and works* (1958), pp. 206–210, and Sauter, Avicennas Bearbeitung [see note 6], pp. 60–63.

[59] On the semantices of "complexio" and "commistio" in medieval translations of medical texts see Danielle Jacquart, "De crasis à complexio: Note sur le vocabulaire du tempérament en Latin médiéval", in: *Centre Jean Palerme, Mémoirs 5, Textes médicaux latins antiques*, ed. by G. Sabah, Saint-Etienne 1984, pp. 71–75.

> most. And because, after all, the first faculties in the aforesaid elements are four, [namely] heat, coldness, moisture and dryness, it is evident that it is through them that the *complexiones* exist in that which has become and that which has ceased to be.[60]

This explicit definition of *complexio* as a quality suggests the possibility that the innate heat, too, could be understood as supra-elementary, complexional, and elemental-qualitative.[61] If, however, the elemental particles with their primary qualities remain integrally preserved in the *complexio*, the question necessarily arises as to how from this mixture a form can be created that is more than a mere rearrangement of the mixed particles.

The presentation of the *complexio* as a kind of sum quality resulting from the primary qualities of the elemental particles undergoes a significant differentiation in the propaedeutic to pharmaceutical teachings with the introduction of a *complexio secunda*:

> The first *complexio* is therefore the *complexio* that comes from the elements. And the second *complexio* is the *complexio* that comes from the things that in themselves have a *complexio*, for example the *complexio* of compound medicines, such as the *complexio* of theriac: after all, each of the individual medicines in the theriac has its own *complexio;* then they are mixed and compounded so that they are united and obtain a *complexio* which is the second *complexio*.[62]

60 Avicenna, Canon, 1556 [see note 1], p. 7 (lib. 1, fen 1, doctr. 3, cap. 1): "Complexio est qualitas, quae ex actione ad inuicem, & passione contrariarum qualitatum in elementis inuentarum: quorum partes ad tantam paruitatem redactae sunt, ut cuiusque earum plurimum contingat, plurimum alterius prouenit. Et quoniam uirtutes primae quae sunt in elementis praedictis, sunt quatuor, caliditas, frigiditas, humiditas, & siccitas: tunc manifestum est quod complexiones in corporibus generatis & corruptis sunt per eas."

61 Jean Fernel's harsh criticism of Avicenna's theory of the "complexio" as a fifth quality, which in some inexplicable way was to emerge from the four primary qualitites (cf. Siraisi, Avicenna [see note 1]) could find support in such passages, but only if the role of God or the "dator formarum" was neglected. Fernel's understanding of celestial heat seems to be quite close to Avicenna's ideas, in fact, but with a stronger vitalist element (see also Léon Figard, *Un médecin philosophe*, Paris 1903).

62 Avicenna, Canon, 1556 [see note 1], pp. 164–165 (lib. 2, fen 2, tract. 1, cap. 1): "Complexio itaque prima, est complexio proueniens ab elementis. Et complexio secunda, est complexio quae prouenit a rebus quae in seipsis habent complexionem: sicut uerbi gratia, complexio medicinarum compositarum, ut complexio theriacae: unaquaeque enim

Such a second *complexio* is said to be found not only in man-made artefacts like pharmaceuticals, but also in nature; for example, it is in this way that milk is composed of wateriness, cheesiness, and butteryness.[63]

Thus, the diverse effects of theriac on a large number of serious diseases and in particular with cases of poisoning are owed not directly to the elemental qualities of the ingredients, but rather to the "property of its form, which follows from the *complexio* of its simple component parts, so that it strengthens the spirit and the innate heat, which is why nature is supported by it beyond the expulsion of the cold and warm [things] that are contrary to it [nature] and overcomes their powers".[64]

On the basis of a particle theory of matter, an understanding of the mixing process emerges here in which the component parts preserve their specific natures, their forms, but at the same time constitute a new form in the first *complexio*. Natural bodies that are endowed with such a first *complexio* can for their part come together to form a second *complexion* in renewed mixture, in which, however, the first *complexiones* and thus the forms of the components of the mixture are preserved. This indicates of a hierarchy of forms. The wording suggests mere sum qualities, but on the other hand, genuinely new properties are created.

This concept of the *complexio secunda* bears a striking resemblance to one of the most powerful concepts of Renaissance medicine, the doctrine of *tota substantia*. According to this doctrine, the specificity of the effects of some medicinal drugs and of some illnesses could not be explained in terms of primary qualities alone, but was rather owed to the "total substance".[65] Scattered references to the concept of *tota substantia* can

medicinarum singularium theriacae complexionem habet sibi propriam, deinde commiscentur & componuntur, ita ut uniantur, & adueniat eis complexio, quae est complexio secunda."

63 Ibid.

64 Ibid., p. 972 (lib. 5, summa 1, tract. 1): "Et nisi efficit has operationes, nisi propter proprietatem formae eius, sequentem complexionem simplicium ipsius, ita ut confortet spiritum, & calorem innatum: quare iuuatur natura per illud super expulsionem contrariantium frigidorum & calidorum, & uincit uirtutes eorum."

65 Cf. Linda Deer Richardson, "The generation of disease. Occult causes and diseases of the total substance, in: *The medical Renaissance in the sixteenth century*, ed. by Andrew Wear et alii, Cambridge 1985, pp. 175–194; the *Canon* uses the term "forma" in the sense of the scholastic *forma specifica* in this context, not the term "tota substantia" (see, for example, Avicenna, Canon, 1556 [see note 1], p. 972 (lib. 5, summa 1, tract. 1); for later developments, see Michael R. McVaugh, "Theriac at Montpellier 1285–1325", in: *Sudhoffs*

already be found in Galen.[66] Yet in the context of Aristotelian-Galenic theory of qualities, the doctrine of the *tota substantia* essentially constitutes a foreign entity, a kind of "black box" to explain the especially mysterious effects of smallest quantities of matter. Considering the great specificity and the fundamental otherness of such effects, it appeared questionable that the form responsible for these effects could be derived in a rationalist manner from the potentiality of matter. As the quotation cited above shows, the description in the *Canon* can be superficially interpreted as an attempt to attribute the supra-elementary faculties, for example of the *forma* of theriac, to the *complexiones* of its component parts. And yet, the doctrine—unmentioned in the depiction of pharmaceutical effects—of a heavenly source of forms, the *colcodea*, opens an alternative interpretive possibility, one that would be clearly contrary, however, to the Galenic claim to rationalist deduction from perceivable qualities, and would thus threaten the very foundations of Galenic philosophy and epistemology.[67]

Much more prominently than in the *Canon*, these tendencies come to the fore in the abovementioned treatise *De medicinis cordialibus*, which

Archiv 56 (1972), pp. 113–144; on the debates about theriac see Thomas Holste, *Der Theriakkrämer. Ein Beitrag zur Frühgeschichte der Arzneimittelwerbung*, Pattensen 1976, pp. 18–39.

66 The clearest definition is in Galen, "De simplicium medicamentorum temperamentis ac facultatibus", in: idem, Opera [see note 14], vol. 11, pp. 704–782, cit. p. 705. On the links between Galen's theory of qualities and his epistemology see Georg Harig, "Verhältnis zwischen den Primar- und Sekundärqualitäten in der theoretischen Pharmakologie Galens", in: *NTM* 10 (1973), pp. 64–81.

67 As a matter of fact, the *Canon* turns out to be the principal source for one of the most important concepts in medieval natural philosophy, the concept of "forma specifica" (cf. Ottosson, Scholastic medicine [see note 32], pp. 265–206; Nancy G. Siraisi, *Taddeo Alderotti and his pupils. Two generations of Italian medical learning*, Princeton 1981, pp. 159–162). The origins of the term are linked to neoplatonic ideas of an external source of the forms.

also seems far more influenced by neoplatonic thought.[68] Discussing the terms *proprietas* and *natura*, it says,

> Whatever the case, it is true that from the forms and the faculties there is a [form] that does not exist in the composition of the simple parts: namely, the preparation of the same is only completed after the simple [parts] become mixed: as for example the iron-attracting faculty is in the magnet, instead of appearing by itself in one of the two faculties in the elements; rather it is a [faculty] that comes from the outside, from divine emanation, appearing in all things and extracting all other faculties from potential to actual, according to one of the two kinds, either through a first preparation in the matter or the *hyle* itself, or through a *complexio*, a *complexio* disposed to receive the aforementioned faculty but that is not this faculty itself and does not bring it about. [...] In summary, then, *proprietas* is the nature that is found in mixed bodies and in them follows from a higher and very copious influence that results from their own *complexio*, which attains a special and individual disposition.[69]

This is a clear rejection of the idea of an "eductio formarum ex potentia materiae".[70] Taking a surprising turn, this rejection then targets

[68] On the impact of neoplatonic ideas on Renaissance and early modern epistemology and on the turn toward experimental approaches and the empirical observation of complex faculties and the growing disenchantment with their rationalist deduction from sensible qualitities see Nelly Tsouyopoulos, "Der Einfluß des Neoplatonismus auf die Wissenschaft der Renaissance", in: *Sudhoffs Archiv* 60 (1976), pp. 33–44; Massimo Luigi Bianchi, "Occulto e manifesto nella medicina del Rinascimento: Jean Fernel e Pietro Severino", in: *Atti e Memorie dell'Accademia Toscana di Scienze e Lettere La Colombaria* 47 (1982), pp. 183–248; and Keith Hutchison, "What happened to occult qualities in the scientific revolution", in: *Isis* 73 (1982), pp. 233–253.

[69] Avicenna, De medicinis cordialibus [see note 53], p. 1046 (tract. 1, cap. 10): "Sed qualitercunque sit, hoc est uerum, quod ex formis & uirtutibus est aliqua quae non existit in dispositione simplicium: & non completur quidem praeparatio ad eam, nisi post commistionem simplicium: uerbi gratia sicut est in magnete uirtus attractiua ferri, & non est inuenta in una duarum uirtutum in elementis per se, sed est ab extra a manatione diuina procedente in omnibus rebus, & extrahente cunctas uirtutes de potentia ad actum secundum unum duorum modorum, aut mediante praeparatione prima inhaerente materiae seu hyle, aut mediante complexione, quae quidem complexio disponit ad receptionem praedictae uirtutis, & non est ipsamet uirtus, nec efficiens ipsam. [...] In summa ergo, proprietas est natura inuenta in commistis corporibus, procedens in eis a superna & plenissima influentia, propter complexionem propriam, quae particularem & propriam dispositionem acquirit."

[70] Nicole Oresme made them his own: "Decipiuntur homines qui omnia composita ex elementis volunt habere et reducere ad virtutes quas primo habent elementa [...]. Multe

a reductionist interpretation of the faculties of even the simplest natural bodies:

> After all, if fire were difficult to find and had to be fetched from far away then certainly most people would be more surprised at its property than at the properties of other bodies: because, indeed, its workings are very curious. To wit, its presence pulls the sense of sight from the potential to the actual, and it does not allow one to touch it, and it makes everything that it is able to overcome rise upward: and within an hour, a modest [measure] of it turns into a great measure of the same, and it destroys everything that it encounters, turning it into its own substance: and it is not reduced when something from it is received. These properties are in truth stranger than the attraction of iron by the magnet or other properties. But because the aforementioned properties are common and appear every day, the marveling and the enthusiasm to search for the causes is missing. The working of the magnet, on the other hand, because it is rare and unusual, gives rise to amazement and encourages an investigation of the causes.[71]

With these ideas in mind, which do not find clear articulation in the *Canon*, however, the nebulous definition of the nature of the *calor innatus* as described above is rendered more readily understandable. Much like fire, with which it has much in common, as the aforementioned lamp metaphor shows, its ostensibly elementary faculties can ultimately only be understood as an expression of a specific nature or form that

sunt forme substantiales et multe qualitates, ut forma asini, equi, arboris, herbe et cetera, complexio [sive] proprietates magnetis, reubarbarii, turbith, tyriace, veneni, et cetera, que non sunt in elementis; de his vide Avicennam in de viribus cordis." (Nicole Oresme, *Contra divinatores*, Bibliothèque nationale, Ms. lat 15126, fol. 30v, cit. in McVaugh, Theriac [see note 65], note on p. 114).

[71] Avicenna, De medicinis cordialibus [see note 53], p. 1046 (tract. 1, cap. 10): "Quòd si ignis esset res difficilis inuentionis, & de longinquis partibus afferretur, admiraretur utique uulgus de proprietate eius amplius quam de causis aliarum proprietatum: quoniam ipsius operationes ualde mirabiles sunt. Etenim praesentia eius extrahit uisum de potentia ad actum, & tangi se prohibet, & ascendit sursum, & ascendere facit ea quae uincere potest: et generatur ex modico ipsius in una hora quantitas magna, & corrumpit omne illud cui obuiat, & transmutat ad sui substantiam: nec etiam minuitur, si accipiatur ex eo. Et haec quidem in ueritate mirabiliora sunt, quam tractus ferri a magnete, uel quam caeterae proprietates. Sed quoniam ista usitata sunt, et quotidie apparentia, cadit ab eis admiratio & solecitudo perquirendi causam ipsorum. Actio uerò magnetis quia rara est, & inusitata, inducit admirationem, & provocat ad inquirendem de cause ipsius."

comes from the outside. Accordingly, these faculties, bound as they are to their substrate, are only incompletely regenerable from food.

Concluding Remarks

The *Canon* clearly underlines the central role of the *calor innatus* as an instrument of the faculties of the soul in the life processes of the human body—in procreation, embryogenesis, growth, and aging processes, and in the organization of the body's ability to heal. Yet with respect to the particular nature of the *calor innatus*, it leaves many questions unanswered. We have seen the ambiguity of the lamp metaphor in the explanation of aging and of hectic fever, the unspecified relation between *calor innatus* and *spiritus*, as well as the difficulty in determining more precisely the character and origin of their forms. Today's reader, engaging in a thorough study of the scattered references and explanations, will be left with hardly any doubt regarding the special character of the form of the *calor innatus* and the great extent to which it is identical with the *spiritus*, which to all appearances entails a heightened emphasis on the material aspect as connected with movement as opposed to the aspect of effective power. However, formulations can also be found that most certainly allow for different interpretations. Time and again we find that it is only through *De medicinis cordialibus* that we can arrive at a clear understanding, which stands in striking contradiction to the traditional Galenic-Aristotelian views, however, particularly with respect to the doctrine of forms.[72]

In this way, our little investigation suggests that we ought to cast some doubt on a widespread appreciation of the *Canon*, one that has deep roots in medico-historical scholarship. A connection has been made between the decline of Salerno, which in the twelfth century was still by far the most important center of Western medicine, and the reception of the Latin version of the *Canon*, which, as a systematic compendium of the entirety of medicine, has been said to have overshadowed everything that had been available up to that point.[73] Yet if one compares it to *Pantegni*,

[72] The differences between Galenic and Aristotelian ideas, by contrast, are repeatedly discussed in the *Canon* already.

[73] Heinrich Schipperges, "Die Rezeption arabisch-griechischer Medizin und ihre Einfluß auf die abendländische Heilkunde", in: *Die Renaissance der Wissenschaften im 12. Jahrhundert*, ed. by Peter Weimar, Zürich 1981, pp. 173–196, esp. p. 182; Lufti M. Sa'di,

it is certainly not readily apparent that the *Canon* is clearly superior for those studying and practicing medicine.[74] The theoretical explanations in the first book of the *Canon* are condensed to the point of severely obstructing comprehension and are, as the discussion of the *calor innatus* and the *humidum radicale* has shown, not free of contradictions. And so, it would take a substantial proviso to characterize them as "a tightly systematic explication of the theories involved, couched in a newly precise terminology".[75]

It seems possible, then, that the *Canon's* undisputed and exceptional attractiveness for learned Western physicians was not so much due to the work's clarity and systematicity, given the obvious need for thorough, at times even sentence-by-sentence commentary—in particular for those parts that enjoyed a great and sustained popularity, namely the physiological chapters of the first book and the opening part of the fourth book, which covers fevers, the most important among the general diseases.[76] What seems decisive, rather, is the work's high theoretical, philosophical standard, which found expression in the aforementioned sections in particular.[77] Behind it was the ideal of the Islamic physician-philosopher, which fell into the open arms of the physicians at the young universities, who were seeking to attain for medicine—which was sometimes still grouped as one of the *artes mechanicae*—an equal status among the sciences and to win for the theoreticians and practitioners of medicine an

Reflection of Arabian medicine at Salerno and Montpellier, in: *Annals of medical history* 5 (1933), pp. 215–225.

74 Constantinus Africanus, "Pantegni", in: Isaac Israeli, *Opera omnia*, Lyon 1515; the philosophical sophistication of the "Pantegni" in comparison to the *Canon* has often been insufficiently appreciated. It emerges not only its broad reception among the most "philosophical" physicians in Salerno, Maurus and Urso, but, for example also in the work of William of Conches (cf. Danielle Jacquart, "Aristotelian thought in Salerno", in: *A history of twelfth-century Western philosophy*, ed. by Peter Dronke, Cambridge 1988, pp. 407–428; Heinrich Schipperges, *Die Assimilation der arabischen Medizin durch das lateinische Mittelalter*, Wiesbaden 1964, pp. 113–118).

75 McVaugh, The "humidum radicale" [see note 26], p. 265.

76 Cf. Siraisi, Avicenna [see note 1].

77 See Charles H. Talbot, "Medicine", in: *Science in the Middle Ages*, ed. by David C. Lindberg. Chicago 1978, pp. 391–428, cit. pp. 401–402.

equal social status.[78] Seen this way, and taking into account the influence of the reception of the "new" Aristotle, the *Canon's* lack of terminological precision and its general need for commentary were certainly not disadvantageous for the up-and-coming university physicians. Rather, the *Canon* posed a welcome opportunity for these medical professionals to furnish proof of their dialectical abilities in working with a text which—more than the *Pantegni* or Johannitius's *Isagoge*—made the necessity of a physician's comprehensive philosophical education abundantly clear.

Not only this, but a certain haziness in the terminology and an ambiguity of the concepts opened the possibility for independent interpretations. This openness was not necessarily a detriment when it came to assuring the enduring relevance of the *Canon* as an unwavering reference point and in terms of the development of speculative physiology in the subsequent centuries. As we have seen, behind an exterior of ostensibly traditional Galenic-Aristotelian terminology, certain heterodox ideas are concealed within the *Canon* and even more so in *De medicinis cordialibus*. Even Daniel Sennert, building on Julius C. Scaliger's work in his efforts to reinterpret and renew ancient atomism, would still lean on Avicenna's definition of a mixture as a maximum approximation of the smallest particles.[79] While one may certainly not exclusively attribute to the *Canon* and Avicenna's understanding of *colcodea* and the *spiritus* in *De medicinis cordialibus* the fact that celestial influences were increasingly being taken into account, to the point that medicine and natural philosophy in the Renaissance posited a constant influx of heavenly heat or heavenly *spiritus*, these works would have offered important reference points for related neoplatonizing views. It is without question not a coincidence that in the case of Marsilio Ficino and Jean Fernel, two of the most original and most widely read physicians of the Renaissance, a pronounced dependency on the *Canon* can be observed, whether this comes in the form of explicit recognition as with Ficino, or in the form of an ostensible philosophical

[78] Cf. Ottosson, Scholastic medicine [see note 32], pp. 68–88; Siraisi, Taddeo Alderotti [see note 67] demonstrates the positive effects of this reorientation on the medical profession on the example of the physicians in Bologna.

[79] Daniel Sennert, *De chymicorum cum aristotelicis et galenicis consensu ac dissensu liber*, 2nd. edn, Wittenberg 1629, p. 210; idem, "Physica hypomnemata", in: idem, *Opera omnia*, Venice 1641, vol. I, p. 23; Julius Caesar Scaliger, *Exotericarum exercitationum liber XV. de subtilitate ad Hieronymum Cardanum*, Hannover 1620.

dissociation, as with Fernel.[80] To express my hypothesis somewhat pointedly, it could be said that what secured the great and enduring success of the *Canon* in Western medicine was not its systematicity and terminological precision, but instead its productive openness and ambiguity, combined with a theoretical-philosophical standard, which, though it was only partly delivered, was extremely attractive to those in the medical profession.

[80] Ficino, De vita [see note 24]; on Ficino's spiritual magic see D. P. Walker, *Spiritual and demonic magic from Ficino to Campanella*, London 1958 (repr. Nendeln 1969]; see also Marion Heitzman, "L'agostinismo avicennizzante e il punto di partenza della filosofia di M. Ficino", in: *Giornale critico della filosofia italiana* 16 (1935), pp. 295–322 and pp. 460–480 and 17 (1936), pp. 1–11; Jean Fernel, *Universa medicina*, Paris 1567, pp. 72–85; Mendelsohn, Heat [see note 9], pp. 24–26; cf. Karl Eduard Rothschuh, "Das System der Physiologie von Jean Fernel (1542) und seine Wurzeln", in: *Verhandlungen des XLX. Internationalen Kongress für Geschichte der Medizin, Basel 1964*, Basel 1966, pp. 529–536; Hoppe, Biologie [see note 9], pp. 98–104; Figard, Un médecin philosophe [see note 63]; Charles Sherrington, *The endeavour of Jean Fernel*, Cambridge 1946. Neoplatonic ideas could find additional support in other works by Avicenna, for example, regarding his theory of spontaneous generation as due to the influx of celestial heat (cf. Bruno Nardi, "Pietro Pomponazzi e la teoria di Avicenna intorno alia generazione spontanea nell'uomo", in: idem, *Studi su Pietro Pomponazzi*, Florence 1965, pp. 305–319).

In Awe of Creation. Daniel Sennert's Conception of Total Substance, Innate Heat, and Spontaneous Generation, and His Atomistic Theory of Form

The Wittenberg professor Daniel Sennert (1572–1637) was one of the most well-known and successful authors of his time. His *Institutiones medicinae*,[1] thanks to its clear organization and compilatory nature, was among the most popular medical textbooks until sometime in the eighteenth century. Particular attention attracted his attempt, in *De chymicorum cum Aristotelicis et Galenicis consensu et dissensu*,[2] to integrate Paracelsian concepts into the conceptual edifice of Galenic

First published in German as "Das Staunen vor der Schöpfung: 'Tota substantia', 'calidum innatum', 'generatio spontanea' und atomistische Formenlehre bei Daniel Sennert", in: *Gesnerus* 50 (1993), pp. 48–65. Translated into American English by Logan Kennedy and Leonhard Unglaub. I am grateful to Eckhard Keßler [1938–2018] for his critical comments on an earlier version of this article.

[1] Daniel Sennert, *Institutiones medicinae*, Wittenberg 1611: further editions came out in Wittenberg 1620 and 1633.

[2] Daniel Sennert, *De chymicorum cum Aristotelicis et Galenicis consensu et dissensu*, Wittenberg 1619; 2nd edn. Wittenberg 1629.

M. Stolberg, *Medicine and the Body in Early Modern Europe*, Palgrave Studies in Medieval and Early Modern Medicine,
https://doi.org/10.1007/978-3-031-84556-7_11

medicine.[3] However, more than many other contemporary physicians, Sennert understood himself as both a physician and a natural philosopher, and he very explicitly cautioned against dismissing natural philosophy as a mere "tirocinium" of medicine.[4] Here, historians have concentrated on Sennert's significance in the history of atomism and in the development of a theory of chemical compounds.[5] But Sennert's explanatory approach was far more comprehensive, and this has yet to be sufficiently appreciated in historical scholarship.[6] Retaining the distinction between the animate and the inanimate and using and reinterpreting traditional medical concepts, he aimed to arrive at a unified interpretation of natural bodies as they appear in both living beings and in inanimate matter. This paper will seek to map out these efforts.

The Theory of Forms and Tota Substantia

The starting point and foundation of Sennert's approach is an attitude of awe, an ultimately religiously inspired reverence toward the diversity and complexity of creation and toward the wondrous and ever specific particularity of things in nature.[7] Against the backdrop of the Aristotelean natural philosophy that prevailed in his time, this position leads him directly into an engagement with one of the central problems of Renaissance philosophy, the interpretation of mixtion or the mixing process.

[3] See Walter Pagel, *Paracelsus. An introduction to philosophical medicine in the era of the Renaissance*, Basel and New York 1958, pp. 333–343; Giancarlo Zanier, *Medicina e filosofia tra '500 e '600*, Milan 1983, esp. pp. 100–103.

[4] Daniel Sennert, *Hypomnemata physica* (first publ. 1636), in: idem, *Opera omnia*, 3 vols, Venice 1641, vol. 1, Proemium.

[5] Kurt Lasswitz, *Geschichte der Atomistik vom Mittelalter bis Newton*, Hamburg and Leipzig 1890, vol. 1, pp. 436–454; Andrew G. van Melsen, *From atomos to atom*, New York 1960; G. B. Stones, "The atomic view of matter in the XVth, XVIth and XVIIth centuries", in: *Isis* 10 (1928), pp. 445–465.

[6] The best overviews to date are those by Tullio Gregory, "Studi sull'atomismo del seicento", in: *Giornale critico della filosofia italiana* 43 (1964), pp. 38–65, and 45 (1966), pp. 44–65, who puts Sennert's theories into the context of atomism in the early sixteenth century, and by Jacques Roger, *Les sciences de la vie dans la pensee française du XVIIIe siècle. La génération des animaux de Descartes à l'Encyclopédie*, Paris 1963, pp. 106–112; see also the entry on Sennert by Hans Kangro in the *Dictionary of scientific biography*, vol. 12, New York 1975, pp. 310–313.

[7] See, for example, the preface to the reader in the *Hypomnemata* of 1641 (no page numbers): "divinae sapientiae infinitos thesauros in rerum Natura absconditos esse".

The fundamental question here is: what is the source of the specific characteristics and faculties of natural bodies; how can something essentially new be generated from the mere mixing of the elements? Sennert strictly rejects the explanation that complex natural entities come to be because the form of that which is becoming is transferred from the potential to the actual. Never could mixing the elements bring forth something other than what was in the elements.[8] At the same time, the accidental spatial order of the smallest particles as in ancient atomism also did not seem to him to provide an adequate explanation for the complexity and regularity in nature. Rather, the higher faculties of complex natural bodies could only have their origin in the corresponding supra-elementary forms, which in no way were imaginable as potential under the dominion of the lower elemental forms. It was his view that the diversity of natural bodies and their faculties corresponded to a diversity of specific, supra-elementary forms.[9]

Sennert's anti-reductionism and his concentration on the dynamic aspect of form as a source of the specific qualities and faculties of natural bodies can be found in his early medical treatises already, in the doctoral theses which he wrote and which his students had to defend. These concerned first and foremost the specific effects of medicinal drugs and the diverse and strange symptoms of contagious diseases, pestilential fevers, and cases of poisoning. Following Jean Fernel, Sennert asked: "Are there illnesses of the whole substance or of the form?"[10] He rejected Fernel's ontological conception of illness. Diseases did not, in his opinion, concern the essential constitution, whereby no *intensio* or *remissio* of the

[8] Daniel Sennert, "Medicina practica", in: idem: *Opera omnia*, Lyon 1656, vol. 3, p. 973: "Mihi si, quod sentio, libere dicendum, vix vllam rei Physicae damnosiorem, & quae optima ingenia irrito & inani labore magis fatigauerit, in Physicam irrepsisse puto opinionem, quam illorum, qui omnium eorum, quae in rebus naturalibus accidunt, caussas e qualitatibus manifestatis, & quatuor elementis reddere conati sunt." In this context, Sennert repeatedly also drew attention to the difficulty of explaining, why the actualization of potentiality occurred at a specific point in time, rather than instantaneously, without any delay.

[9] Sennert, De chymicorum, 1619 [see note 2], p. 363: "Non tamen [...] ex fortuito atomorum concursu res naturales fiunt [...], sed a forma superiore dirigente, quae Caloris instrumento & ope attrahit, retinet, miscet, & disponit omnia".

[10] Daniel Sennnert, *Quaestiones medicae controversae septem*. Resp. Michael Döring, Wittenberg 1607, Quaestio secunda.

qualities was possible, but rather they were accidental.[11] He did, however, observe in the experiential world many examples of special occult effects of certain substances on the human body, which could not have their origin in the forms of the elements but only in higher supra-elementary forms the *tota substantia*. There were certain diseases that corresponded to them and which could thus be characterized as diseases of the whole substance.[12] He discussed these matters with reference to the purgative effects of cathartics, for example.[13] Their ability to target and eliminate a harmful humor from the body was not, in his view, adequately explained by the attractive faculty of elemental heat. There must be a specific attractive faculty that could result from a higher specific form alone, from the whole substance and its relation to the humor in question. He was adamant in his defense against the accusation that he was simply disguising his ignorance when he posited occult qualities and a whole substance. It was rather the case that those men were guilty of foolhardiness who tried to penetrate the deepest secrets of nature and put manifest false qualities in the place of hidden but true ones.[14] Experience, he wrote, showed that "all things have been endowed with certain higher qualities through which they bring about by means of the nature and their immanent movement of the form what is proper to them".[15]

[11] Ibid., p. 13: "Essentia autem rei intendi atque remitti non potest."

[12] Ibid.: "ita [in rebus naturalibus, M.S.] non solum qualitates illae manifestae, caliditas, frigiditas, humiditas, siccitas, & si quae alia sunt, reperiuntur, sed etiam nobiliores quaedam & occultiores, quae à formis illis nobilioribus fluunt & emanant, atque a quibus nobiliores & praestantiores actiones proficiscuntur. Atque ut illae mutuo se afficiunt: ita etiam inter occultas actionem quandam esse, negandum non est: cum ipsa id testetur experientia; quam quidem occultarum qualitatum actionem vulgo a tota substantia nominant." On the concept of "tota substantia" and on Sennert's disease theory see Linda Deer Richardson, "The generation of disease. Occult causes and diseases of the total substance", in: *The medical Renaissance of the sixteenth century*, ed. by Andrew Wear, Roger French, and Iain M. Lonie, Cambridge 1985, pp. 175–194, and Peter H. Niebyl, "Sennert, van Helmont and medical ontology", in: *Bulletin of the history of medicine* 45 (1971), pp. 115–137.

[13] Daniel Sennert, *Quaestiones medicae controversae quinque*. Resp. Ioachimus Koppe. Wittenberg 1607, Quaestio tertia: "Qua ratione pharmaca kathartika [graece] humores expurgent?".

[14] Ibid., Quaestio tertia, 26: "temerarij esse judicamus, rerum abditissimarum arcana perscrutari, & in naturam injurij pro veris occultis falsas manifestas reponere velle."

[15] Ibid., Quaestio tertia, 25: "rebus omnibus proprietates quasdam nobiliores inditas esse, quibus quod sibi proprium est agunt, pro forma natura & motione insita [...]."

Sennert further developed and specified these ideas in his *Institutiones medicinae* of 1611.[16] Citing Galen, he introduced an essential argument—one that he would subsequently repeat many times—that supported the existence of supra-elementary qualities and corresponding forms: their ability to achieve the greatest effects independently of their mass and even in the smallest quantities, which distinguished them from the elemental qualities.[17] The quantity of poison emitted by a scorpion was so small that one could not even see a hole in the animal's stinger, and yet the consequences were disastrous.[18] The same was true of contagia:

> what is passed on is something unnatural [...] a certain body, a seed as it were [...] But it is part of the nature of this seed that it contains that occult quality that can bring forth in the body in which it is taken up a so-called disease of the whole substance: which is why a tiny quantity holds a great power of the contagium. While the elemental qualities, so as to have a strong effect, require much matter, these [occult] qualities, as they are partisans of form, have great power even in a tiny quantity, and a tiny quantity of such seed is enough to infect the entire body.[19]

Sennert found further evidence of the existence and mode of action of supra-elementary qualities in the specific sympathetic relationships, for example, as they become discernable in the case of contagia infecting

Similar ideas were already proposed by Jakob Schegk, *Tractationum physicarum et medicarum tomus unus*, Frankfurt 1585, who used the traditional terminology, writing of "forma specifica" and effects "kat'entelecheian".

[16] In Sennert, Institutiones, 1611 [see note 1], pp. 131–145, chapter "De morbis totius substantiae seu occultarum qualitatum", Sennert discussed the concept of "tota substantia" in even greater detail.

[17] Ibid., p. 242: "Neque enim qualitates primae in minimo corpore summas vires exercere possunt." Sennert quoted Galen, *De simplicium medicamentorum facultatibus*, book 5, ch. 19: "Sunt quae tota substantia nobis sunt contraria. [...] Nam talia omnia genere sint deleteria, non quantitate."

[18] Sennert, Institutiones, 1611 [see note 1], p. 242.

[19] Ibid., p. 248: "quod communicatur, est quiddam praeter naturam, [...] corpus quoddam & semen quasi [...]. Ad hujus autem seminis naturam pertinet, ut qualitatem illam occultam, qua morbum a tota substantia dictum in corpore, a quo recipitur, excitare possit, obtineat: ob quam etiam in exiguo corpore magna contagii vis est. Cum enim elementares qualitates, ut multum agant, etiam multam desiderant materiam: hae qualitates, quia formae asseclae sint, etiam in exiguo quanto magnam habent vim, atque ad inficiendum totum corpus minima quantitas talis seminis sufficit."

individuals in different ways, and with certain poisons and remedies in the way they target certain body parts.[20]

Calidum Innatum

The attempt to ascribe specific faculties, especially those of medicinal drugs, to a *tota substantia* in the sense of a supra-elementary form leads directly to the question of the material basis of these occult qualities and to the question of their relationship to the souls of the plants or animals from which many drugs are derived. Sennert clarifies that the form in question, which is effective in the drug by means of its occult qualities, can by no means be identical with the soul of the plant or animal, which is by definition absent in dead plants and animals.[21]

However, if this supra-elementary form cannot be owed to the mere *potentia* of the elemental matter that remains after death, then where does it come from? This question ultimately concerns not only medicinal drugs, but all natural bodies with supra-elementary forms and faculties. Sennert decidedly rejects the neoplatonizing assumption of an influx of form from an external source. He can only imagine divine intervention in creation to occur as a miracle. Due to his immanentist understanding of nature, it seems, he did not find the doctrine of the continual work of an *anima mundi*, a *colchodea*[22] or any other kind of heavenly giver of form to be acceptable. This leaves essentially only one possibility, however. The form in question must have prexisted *in actu* in the living body before this body died, but without being identical with the form of the deceased body—that is, in the case of plants and animals, with the soul.

Within the framework of the traditional Aristotelian theory of forms, this seems to make no sense.[23] Since the form determines the essence, the suchness of the respective natural body, two different forms cannot be responsible for it at the same time. A major merit of Aristotle's theory of

[20] Ibid., pp. 250–252 and p. 807.

[21] Sennert, De chymicorum, 1619 [see note 2], pp. 248–249; idem, Hypomnemata, 1641 [see note 4], p. 14.

[22] On the "colchodea" see Bruno Nardi, "Origine dell'anima umana", in: idem, *Studi su Pietro Pomponazzi*, Florence 1965, pp. 231–246, cit. note on pp. 234–245.

[23] For a survey of Aristotle's natural philosophy, see Friedrich Solmsen, *Aristotle's system of the physical world*, Ithaca and New York 1960; Ingmar Düring, *Aristoteles. Darstellung und Interpretation seines Denkens*, Heidelberg 1966.

actuality and potentiality lay precisely in the fact that it offered a solution to this dilemma. Sennert, however, eludes these difficulties by adopting the imagery of hierarchical political relations of rule and constructing a hierarchy of forms within the complex cosmos of the things of nature and living beings.[24] The highest, ruling form—the soul in the case of animals and plants—makes use of the subordinate forms for its purposes. These are "taken for themselves, also true forms, which form their matter, but in their relationship to the soul and the specific form they serve immediate matter".[25] When, however, the ruling form ceases to be and the unity it has brought forth disintegrates, "these forms continue to exist for some time and act as forms and are the originators of the faculties and functions that are found in many no-longer-living things and nevertheless cannot arise from the qualities or forms of the elements".[26]

It is at this point in the argumentation that Sennert takes recourse to one of the oldest physiological concepts of Western medicine: the doctrine of implanted or inborn heat and the concepts of a *spiritus innatus* and a *humidum radicale*, that were closely connected to it.[27] The diverse faculties which the inborn heat manifested as an instrument of the soul in the processes of nutrition and procreation made it clear that its work could not be attributed to the elemental qualities.[28] For Sennert, this heat is not

[24] Sennert (Hypomnemata, 1641 [see note 4]) is drawing on Jacopo Zabarella's theory here, according to which the form of the mixed assumed the "officium" of a specific form, when the soul perished, to which the mixed had previously served as matter; the passage is in Jacopo Zabarella, *De rebus naturalibus libri XXX*, 4th edn, Cologne 1602, pp. 395–399 (liber de communi rerum generatione).

[25] Sennert, Hypomnemata, 1641 [see note 4], p. 14: "Quae quidem formae per se consideratae etiam verae formae sunt, quae materias suas informant: sed respectu animae & formae specificae rationem materiae immediatae habent."

[26] Ibid.: "remanent aliquandiu hae formae, & formae officio funguntur, & facultatum atque operationum, quae in rebus pluribus amplius non viventibus deprehenduntur, neque tamen ex elementorum qualitatibus vel formis provenire possunt, authores sunt."

[27] Daniel Sennert, *Epitome naturalis scientiae*, Wittenberg 1618, p. 605: "Constat autem calidum innatum spiritu insito & humido primigenio: estque calidum innatum humidum primigenium insito spiritu & calore undique perfusum." On the history of the concept of "calidum innatum" see Everett Mendelsohn, *Heat and life. The development of the theory of animal heat*, Cambridge 1964; Friedrich Solmsen, "The vital heat, the inborn pneuma, and the aether", in: *Journal of Hellenic Studies* 77 (1957), pp. 119–123.

[28] Sennert, De chymicorum, 1619 [see note 2], pp. 247–248: "Etsi ignis in mistis utpote temperatus, alio modo agat, quam simplex: tamen nec tum ultra suas vires agit.

merely a faculty in the service of the soul. Rather, it is matter, a *calidum*.[29] As such, however, the concept of the *calidum innatum* serves as an excellent key to understanding the effects of the medications mentioned above. After the death of the plant or animal, and thus after the related loss of dominion of the soul, the *calidum innatum*—the living thing's heretofore materially based instrument—takes over in Sennert's account. It is the source and material substrate of supra-elementary healing faculties.

Here Sennert, at least in passing, extends the doctrine of the *calidum innatum* far beyond the realm of physiology. To him, the effective power of medicines appears to point in large part to specific relations of sympathy and antipathy. It therefore plausible that the *calidum innatum*, or something analogous to the *calidum innatum* in the inanimate, must be seen as the very source of the relations of sympathy and antipathy that permeate Sennert's universe.[30]

Sennert sees the things of nature as shot through with something "spirituosum" which goes beyond the elementary.[31] It is called "calidum innatum" only due to its noticeable heat in living things. For him "not only the heat is the tool of the soul, but the entire matter that is called by the name of implanted heat or 'spirit' and which is endowed with not only heat but with other higher qualities as well".[32] To gems, stones, and minerals in particular Sennert ascribes with such a spiritus, analogous to the *calidum innatum*.[33]

Et nihil praestat nisi quod in sua & reliquorum, cum quibus mistus est, elementorum potestate est."

[29] Sennert, Epitome [see note 27], p. 605.

[30] Sennert had discussed this topic already as a student in a disputation: see Johann Jessenius von Jessen, *De sympathiae et antipathiae rerum naturalium caussis disquisitio singularis, quam in publico pro virili ad Cal. Iunij defendere conabitur M. Daniel Sennertvs Vratislaviensis*, Wittenberg 1599.

[31] Sennert, De chymicorum, 1619 [see note 2], p. 260: "cum multis doctis viris, Fernelio, Valeriola, Schegkio, & alijs plurimis probabiliter asserimus: 'Esse quid spirituosum in rebus naturalibus praeter elementa'". Sennert did not neglect to point at the Aristotelian characterization of "spiritus" as "aitheroides", that is, as similar to the matter of the stars (Sennert, Hypomnemata [see note 4], p. 8).

[32] Sennert, Institutiones, 1620 [see note 1], pp. 35–36: "Neque sola caliditas est instrumentum animae: sed totum illud corpus, quod calidi vel spiritus insiti nomine venit, & non solum calore, sed & aliis nobilioribus qualitatibus instructum est."

[33] Sennert, De chymicorum, 1619 [see note 2], pp. 257–259. Later, Sennert used the terms "spiritus" and "calidum" more reluctantly again and attributed the occult qualities of gems, stones, and minerals more generally to their supraelementary, specific form; even

Sennert finds crucial evidence of the existence of special, supra-elementary matter in the teachings and laboratory experience of the alchemists to which he frequently refers. They succeed in isolating the effective healing part of plant or animal matter, separating it from the ineffective parts through distillation, filtration and other separation methods and thus rendering the *calidum innatum* visible[34]: "When namely, through concoction or decoction, the part of the substance containing the purgative power is separated and extracted, a useful medication is prepared; the remaining the matter, however, stays behind and is entirely ineffective".

The question remains, however,—especially for the inanimate—how this materially anchored supra-elementary form with its amazing faculties has found entrance into the natural body in the first place so as to be able to take over dominion, once the specific form of the body ceases to be. As with his position in the doctrine of forms, Sennert, opposing Jean Fernel, strictly rejects the idea of a constant influx of supramundane heat or spiritus. For him, the *calidum innatum* does not flow into the *kyema* at the moment of conception, but rather is passed on like the soul from generation to generation and thus ultimately goes back to the divine creation of the world. The imagery of procreation, however, also becomes paradigmatic in the explanation of the origin of supra-elementary forms and faculties in inanimate things of nature. In the biblical creation story, Sennert recognizes an important advance in knowledge in the Christian occident that supersedes the Aristotelian teaching of the eternity of the world: "More correctly, the holy scriptures teach us that all things have their forms – through which they are distinguished from one another and are effective – from the first act of creation, and they retain these forms still today through the will of the almighty God".[35] "This is to say that God placed a seed of every thing as a rudiment for the becoming of things. Through it, that which is capable of becoming procreates. He

plants, he later wrote in the *Hypomnemata* (Sennert, Hypomnemata 1641 [see note 4], pp. 13–14) were endowed only with something analogous to the "calidum innatum".

[34] Sennert, Hypomnemata 1641 [see note 4], p. 14: "Si enim coctione vel infusione ea substantiae pars, in qua vis purgandi est, separetur & extrahetur, vtile medicamentum paratur; reliquum vero corpus plane inefficax relinquitur."

[35] Ibid., p. 4: "Rectius, vt dictum, sacrae literae nos docent, in principio elementa creata esse, & res omnes [...] formas, quibus a se inuicem differunt & efficaces sunt, ex prima creatione habere, quasque Dei Opt. Max. voluntate ad hunc isque diem conseruant."

himself mixed the whole together at the beginning, and so the body, which was given to things at the beginning, is passed on in becoming and is constituted, ruled and preserved to this day by those principles that were established in the beginning".[36]

The specificity of individual things as well as their faculties thus do not originate in a mixing of elements or in an accidental joining of atoms, but ultimately in the divine act of creation. In Sennert's view, the forms along with their proper substrates have existed since the beginning of the world.

In this way, the traditional problem of *mistio* is literally turned on its head. The form of that which is becoming is no longer the result but rather the cause of the mixing process and its specific course of events. Proceeding from small material *quanta*, it subordinates the surrounding matter to its domain, guiding and organizing. The diverse forms of the things of nature are thus not new creations but rather have always existed, and they propagate, if an analogy can be made at all, in the way of a candle's flame, which can be passed on to another candle without the first going out.

Generatio Spontanea

This account had far-reaching consequences for the fundamental and much-discussed problem of natural philosophy to which Sennert devoted a large part of his final work, the Hypomnemata physica[37]: the question of the nature of spontaneous generation.[38] Corresponding to his anti-reductionist stance, Sennert strictly rejects attempts to trace the generation of worms and insects from rotten flesh and similar generally recognized phenomena to a mere "eductio formarum ex potentia materiae". For him, it seems impossible to reduce the many different faculties and complicated structures of even simple, spontaneously generated living

[36] Sennert, De chymicorum, 1619 [see note 2], p. 261: "Creator enim rerum generationis fundamentum posuit seminium cujusq[ue] rei, per quod generabilia propagantur, ac ipse primitus totum miscuit; atque ita in rerum generatione hoc, quod ab initio rebus inditum corpus, in generatione propagatur, & ab ijs, quae primo condita sunt, principijs adhuc hodie constituitur, regitur & conservatur."

[37] Sennert, Hypomnemata 1641 [see note 4], pp. 65–97.

[38] On the early of this concept see Erna Lesky, *Die Zeugungs- und Vererbungslehren der Antike und ihr Nachwirken*, Mainz 1950.

things to the potential of elemental matter. He holds that in some cases the generation is only apparently spontaneous because, unbeknownst to the observer, veritable seed matter of the respective species that is being generated has gained entrance.[39] Otherwise, when an external source of forms is ruled out, recourse must be made to a solution that is quite similar to the explanation of the medicinal effects of plant and animal substances. The form of the new, spontaneously generated living thing must have been present during the lifetime of the being which has now died but which was previously animated with a higher form.[40]

Again, the question arises here, and more urgently than in the discussion of occult qualities and their origin, how not only several forms, but even of several souls could coexist in one living thing. In his interpretation of spontaneious generation, Sennert therefore feels called upon to further differentiate his conception of a hierarchy of forms in living and non-living natural bodies. Based on the authority of Aristotle, he states that all things are in some sense full of soul.[41] This is not to be understood, however, in the sense that all things of nature are animated, but Aristotle wanted to say "that there is in almost all things a substance of approximately this quality, and when it meets with suitable matter and is free of all hindrances, it reveals itself and works in the office of the soul".[42]

The soul of the spontaneously generated living thing has certainly been, in his account, *in actu* in the previous living thing, only not as a specific, but as an ancillary, subordinate form. For this reason, one must not think that human flesh holds actual worms but rather only the forms of worms.[43] He suggests once more that he sees this for now still

[39] Sennert, Hypomnemata 1641 [see note 4], p. 72.

[40] Sennert left it open, however, how worms, as animals, which should have been endowed with a higher form than plants, could be spontaneously generated from the wood of a plant.

[41] Sennert, Hypomnemata 1641 [see note 4], p. 71: "quoddammodo animarum plena sunt omnia".

[42] Ibid., p. 71: "quod in omnibus fere rebus sit talis fere substantia, quae, vbi omnibus impedimentis remotis idoneam materiam nacta est, sese exserit, & animae officio fungitur." Sennert also refers to his favorite authority here, Julius Caesar Scaliger, who had attributed spontaneous generation to a seed-like matter (ibid., 73). On Scaliger, cf. Van Melsen [see note5], pp. 73–77.

[43] Sennert, Hypomnemata 1641 [see note 4], p. 73.

subordinate form as rooted in the *calidum innatum*,[44] yet unlike in *De chymicorum* the question of the nature of the form takes precedence over the possible material substrate.

Considered closely, in other words, spontaneous generation no longer corresponds at all in Sennert's account with the becoming of a new living thing that has a new form, but instead describes an accession to rule by a form that has existed for a long time, whose faculties up to that point have simply been in the service of the higher form. The new form is just as old as the one that is ceasing to be, the origin of both being the divine act of creation. This essentially abolishes the special natural-philosophical status of spontaneous generation. In the end, it follows the same laws as sexual procreation: a seed-like matter subjects the matter that surrounds it and communicates itself through its form. It is only logically consistent that Sennert credits the same matter with the ability to then procreate sexually again afterward.

Atomism

It has come to light in what has been discussed so far that Sennert's assumption of a corpuscular structure of matter had various merits which helped explain specific natural phenomena. It allowed all those phenomena to be understood in which tiny or even invisible quantities of matter were present that nevertheless retained their complex constitutive faculties. This was true for poisons, medicinal drugs that had been reduced to the smallest pieces, and contagia, and for the tiniest particles of seed in the case of merely apparent spontaneous generation, or else for laxatives and the transfer of their powers through breastmilk and altogether for the components of food and their distribution in and assimilation by the body.[45] It offered a conclusive explanation for how specific characteristics in different physical states were preserved, for example, when substances dissolved in a solution were passed through a filter of the finest paper and afterward solidified. Hypothesizing a particle structure also helped explain those processes that we would today characterize

[44] Ibid., p. 76: "anima subiectum illud habet, in quo erat in corpore vivente, dum superiori formae loco materia proxima esset."

[45] Ibid., pp. 20–21.

as changes in the state of matter.[46] No longer was the evaporation or vaporization of water understood in an Aristotelian sense as a conversion of the element of water into air,[47] which would have once again raised the question of the origin of the form of air, but analogously as a thickening or thinning of elemental particles, which retained their form despite the very different volumes of space they occupied.[48]

Sennert's concept, in other words, according to which natural bodies are built from tiny parts that are ruled by specific elemental and supra-elementary forms, proves to be quite powerful in explaining a host of phenomena. He was furthermore able to find corroboration in the repeatedly cited definition of "complexio" in Avicenna's *Canon medicinae*, thus relying on a medical authority to which Julius Caesar Scaliger had already taken recourse. According to this definition, *complexio* results when the smallest particles come into maximum contact with one another and their qualities act on each other.[49] The phenomena that serve as examples in turn offer Sennert a proof of the validity of his explanatory approach. He is particularly fond of citing the example of chalk deposits in thermal baths as illustrating that the smallest particles could exist even in the clearest water.[50]

The existence of a corpuscular theory does not imply the assumption of true atoms, of indivisible minima of matter, however, and this restriction on divisibility is a central characteristic of Sennert's account of particles. If he had at first rejected Democritean atomism as reductionistic, now he turned against Aristotle, who had in his criticism of Democritos obscured the fundamental distinction between the mathematical bodies that are

[46] Ibid., p. 21.

[47] Sennert himself had still shared this view in 1618 (Sennert, Epitome [see note 27], p. 232).

[48] Sennert, Hypomnemata 1641 [see note 4], p. 76.

[49] Sennert (Institutiones, 1611 [see note 1], p. 15) quotes Avicenna's definition of "complexio" as the "qualitas, quae ex mutua actione & passione contrariarum qualitatum elementorum quorum partes ad minimas redactae, ut cuiusque earum plurimum contingat plurimum alterius, provenit."

[50] Sennert, De chymicorum, 1619 [see note 2], p. 362. Long before he developed his atomism, he had explained, in an analogous way, the generation of kidney- and bladder-stones inside the body as resulting from the gradual deposition and accretion of tiny salt particles, which could not be perceived as long as they were mixed with the bodily fluids (Daniel Sennert, *Quaestiones medicae controversae quinque*. Resp. Donatus Freywaldus, Wittenberg 1608, Quaestio secunda).

truly infinitely divisible, and the natural bodies that are only somewhat divisible.[51] Democritos, he writes, had concerned himself solely with the latter when he searched for the smallest, indivisible particles from whose congress the natural bodies are generated.[52]

In the chapter "De atomis" in the third book of his *Hypomnemata*, Sennert developed a detailed system of the different kinds of atoms.[53] First, there are the atoms of the elements. Due to their smallness and ability to penetrate, they allow for an entirely unorthodox interpretation of the interaction between the primary qualities. The alteration in the qualities of a natural body is no longer explained in a traditional Aristotelian manner in terms of the mutual *actio* and *passio* of the qualities, but rather as the result of a local motion. It results from a penetration by elemental particles, which consequently contribute to the body with their respective qualities.[54] Thus, the heating of water corresponds to the entering of fire particles, while its cooling corresponds with their exiting.

Sennert locates the *prima mista* on a level above that of elemental atoms. They dissolved and recombined anew during fermentation, digestion and concoction and, by implication, also in alchemical procedures. Situated above the *prima mista*, finally, are the atoms of animated bodies, which at times even contain a complete soul, which was crucial for Sennert's interpretation of spontaneous generation.

All these different atoms, Sennert argued, share the fundamental capacity of retaining their integral form. To preserve their form, natural bodies, however, require a certain minimum of matter that is determined by nature.[55] Strictly speaking, this does not preclude a further division of atoms. If they were divided over a certain minimum limit, they would lose their form, yet the forms that had so far been subordinate could, at least in the case of supra-elementary forms, assume dominion over the resulting

[51] Sennert, De chymicorum, 1619 [see note 2], p. 359; idem, Hypomnemata, 1641 [see note 4], p. 17.

[52] Sennert, Hypomnemata, 1641 [see note 4], p. 17.

[53] Ibid., pp. 16–22.

[54] Ibid., p. 19: "alterationes vulgo dictas non esse saltem in qualitatibus mutationes, sed corporis alieni participationem".

[55] Ibid., p. 22: "Ita enim cum rebus naturalibus comparatum est, vt formae non nisi certo quanto consistant; & sicut vltra naturales terminos sese non diffundere: ita non nisi in certo quanto conseruari possunt."

smaller *quanta* of matter. In other words, Sennert's interpretation can best be understood as a theory of form minima.

This does not yet satisfactorily explain, however, what drove Sennert to assume such true minima. One reason might be found in his involvement in the discussion about the *minima naturalia*, that is the question of whether for natural bodies there is a limit to divisibility below which they cannot exist.[56] He repeatedly quotes the *Optica* of 1613 by Aguilonius, who postulated that there were real minima of light below which light was no longer further weakened but was instead extinguished entirely.[57] The same also went, Sennert explained, for qualities and faculties and what is more, for the forms themselves. Below a certain limit, they would have to cease to exist, not because of the influence of opposing qualities, but rather simply because of the absence of a minimum of material mass—a prerequisite determined by the laws of nature that was indispensable to their self-preservation.[58]

Sennert's atomism thus stands in a centuries-old tradition, and in fact he expressed astonishment over the insinuation that he was proposing something new with his atomism.[59] A close analysis, however, produces a further reason, unacknowledged by Sennert, why he might have turned to a true minima theory as opposed to a mere particle theory: after the Groningen professor Johannes Freitag launched his now famous accusation of heresy against him, Sennert would have felt a pressing need to defend himself.[60] Freitag accused Sennert of postulating the immortality of animal souls and therefore of calling into question the singularity of

[56] See the magisterial analysis in Anneliese Maier, *Studien zur Naturphilosophie der Spätscholastik*, Rome 1949, pp. 155–188.

[57] Franciscus Aguilonius, *Opticorum libri sex philosophis iuxta ac mathematicis utiles*, Antwerp 1613, p. 379 (book 5, proposition 8).

[58] Sennert, Hypomnemata, 1641 [see note 4], p. 22: "Quo pacto & corporum naturalium minima dantur, quae nimirum, si amplius dividuntur, formam essentiamque deperdunt." According to Sennert, this was not due to the action of contrary qualities, "sed solius mensurae defectu, quae ad propriam cuiusque tutelam velut necessaria conditio irrefragabili naturae lege expositur."

[59] Ibid., p. 17.

[60] Pierre Bayle, *Dictionnaire historique et critique*, Paris 1820–1824 (repr. Geneva 1969); see also Wolfgang U. Eckart, "Der Streit zwischen Daniel Sennert (1572–1637) und Johann Freitag (1581–1641)", in: *Deutsch-Niederländische Medizinhistorikertreffen*, Münster 1978, pp. 21–35.

the human soul. Sennert asked leading theological faculties for their judgment on the matter and they exonerated him from this accusation.[61] Yet Freitag was not entirely in the wrong, at least as long as Sennert did not explicitly postulate true minima for matter endowed with form.

If the soul as the form of the animal could have endured in any quantum, no matter how small, this indeed would have given rise to the question as to how such a soul could perish at all. Domination by a higher form was more or less out of the question for the high-ranking animal souls, and in any case, it would have been equivalent not to a annihilation but only to a subordination of the animal soul in Sennert's account. The powerful impact of primary qualities, in turn, would rob the soul of its necessary instruments and would prevent its actualization of all faculties, but it was not equivalent in the framework of Sennert's corpuscular theory of mixtion to annihilation of form and its specific matter. Even if the animal soul left the dying body with a spiritual something as its specific matter, it would continue to exist, even if it did not give shape to a new living thing. In other words, the fundamental difference between matter-bound animal souls and immortal human souls, which were unique in that they were not necessarily bound to matter, would have been eliminated. The animal's soul, too, would be immortal. It was only when the specific matter of the soul, the *calidum innatum*, was broken up so much that a natural limit was reached below which the animal's soul could no longer persist, that this soul would truly have to perish. It is here that Sennert's assumption of a quantitatively defined minimum of form attains a central meaning. It offered no solution for other aspects of Sennert's natural philosophy, however, which had implications that were at least equally controversial from a theological perspective, such as his strictly immanentist theory of form and his assumption, which Freitag did not address, that the individual human soul was materially transmitted and multiplied over generations.[62]

[61] Daniel Sennert, "De origine et natura animarum in brutis", in: idem, *Opera omnia*, 3 vols, Lyon 1676, vol. 1, pp. 285–306.

[62] Soon after its publication, the *Hypomnemata* was put on the *Index of Forbidden Books*. Pietro Redondi has suggested that this may have been due to the theological implications which Sennert's atomistic theory of matter had for the interpretation of the transsubstantiation of the Eucharist, which Redondi also deems decisive for the verdict on Galileo Galilei (Pietro Redondi, *Galileo der Ketzer*, Munich 1990).

Conclusion

It can be said in summary that Sennert furnished an impressive, carefully thought-out and in many respects original natural-philosophical theory. Certainly, his way of thinking no longer accords with the changed world-view we hold today, although if we take a positivistic progress-oriented view, some of his theses strike us as fortunate intuitions, for example his assumption of specific—but unchangeable—matter that is passed on over generations and is endowed with form which in embryogenesis subordinates the surrounding matter and re-orders it to suit its purposes. Sennert's basic scientific and epistemological stance, however, appears to be even more significant.[63] Here it comes to light once again just how diverse and conceptually proficient Aristotelizing natural philosophy was in the late Renaissance. This is seen in his emphasis on the dynamic character of the forms and their knowability on the basis not of intellectually deduced essence but of observable qualities and faculties; in his unperturbed approach to the theorems of the classical authorities, including Aristotle, whom he admired greatly; and in his anti-reductionism which guides people to an empirical study of the ever specific peculiarities and faculties of things as given by the laws of nature and as solidly anchored in matter. His influence on subsequent developments in philosophy—on Leibniz in particular—remains to be studied more closely. In the interpretation Sennert gives it, the application of the form-matter doctrine is far from demonstrating theoretical ossification. If anything, it reflects a recognition of the complexity of nature, which is ultimately traced back to the unfathomable divine act of creation, and to the interaction of things in the context of the laws that govern them and which are the object of human study.

[63] Some suggestions in Wolfgang U. Eckart, *Grundlagen des medizinisch-wissenschaftlichen Erkennens bei Daniel Sennert (1572–1637) untersucht an seiner Schrift 'De chymicorum ... liber', Wittenberg 1629*. Diss. med., Münster 1978.

Index

M. Stolberg, *Medicine and the Body in Early Modern Europe*, Palgrave Studies in Medieval and Early Modern Medicine,
https://doi.org/10.1007/978-3-031-84556-7

FSC
www.fsc.org
MIX
Papier | Fördert
gute Waldnutzung
FSC® C083411

Kolibri 360 GmbH
Ferdinand-Jühlke-Straße 7
99095 Erfurt, Deutschland
produktsicherheit@kolibri360.de